Enhancing Building Resilience Under Climate Change

Enhancing Building Resilience Under Climate Change

Guest Editors

Chengliang Fan
Yundan Liao

MDPI
Basel • Beijing • Wuhan • Barcelona • Belgrade • Novi Sad • Cluj • Manchester

Guest Editors
Chengliang Fan
College of Architecture and Urban Planning
Guangzhou University
Guangzhou
China

Yundan Liao
School of Civil Engineering and Transportation
Guangzhou University
Guangzhou
China

Editorial Office
MDPI AG
Grosspeteranlage 5
4052 Basel, Switzerland

This is a reprint of the Special Issue, published open access by the journal *Buildings* (ISSN 2075-5309), freely accessible at: https://www.mdpi.com/journal/buildings/special_issues/X8VW526VJR.

For citation purposes, cite each article independently as indicated on the article page online and as indicated below:

Lastname, A.A.; Lastname, B.B. Article Title. *Journal Name* **Year**, *Volume Number*, Page Range.

ISBN 978-3-7258-7901-4 (Hbk)
ISBN 978-3-7258-7902-1 (PDF)
https://doi.org/10.3390/books978-3-7258-7902-1

Contents

About the Editors

Chengliang Fan

Chengliang Fan is currently an Assistant Professor at the School of Architecture and Urban Planning, Guangzhou University, and once served as a visiting scholar at the University of Colorado, USA. His main research interests focus on resilient cities, green and low-carbon cities, and urban heat islands, and he has long been committed to theoretical research and practical exploration in related fields. In his scientific research work, he has presided over 6 national, provincial and municipal-level vertical projects, and participated in 3 major projects and general projects of the National Natural Science Foundation of China. With solid research accumulation, he has published more than ten papers at home and abroad. In addition, he also serves as a reviewer for several internationally renowned journals, contributing to academic exchange and quality control in the field, and has successfully developed the "Urban Heat Risk Assessment System" software platform for urban environmental assessment and urban planning, providing a practical tool for urban environmental governance and planning practice.

Yundan Liao

Yundan Liao graduated from City University of Hong Kong and is currently the Deputy Director of the Department of Building Services Engineering, School of Civil Engineering, Guangzhou University. She has presided over 2 National Natural Science Foundation projects and 3 provincial and municipal-level projects; in addition, she has participated in more than 10 projects, including the National Natural Science Foundation projects and provincial-level projects. She has published 30 research papers, participated in the compilation of 1 industry group standard, and won many awards such as the China Patent Excellence Award, the Guangdong Provincial Science and Technology Progress Award, and the Outstanding Academic Award of City University of Hong Kong.

Preface

This Reprint focuses on building climate resilience amid escalating climate change impacts—rising temperatures, extreme weather, and shifting precipitation—that threaten building safety, durability, and functionality. Its scope covers building structure, energy efficiency, performance simulation, water management, green infrastructure, and nature-based solutions. Aimed at providing actionable insights, its purpose is to guide the integration of resilient design and innovative technologies to create buildings that withstand and thrive in climate hazards. Motivated by the urgent need to address climate-related risks and ensure urban/rural sustainability, it is addressed to architects, engineers, policymakers, urban planners, and communities invested in building resilient, livable built environments.

Chengliang Fan and Yundan Liao
Guest Editors

buildings

MDPI

Article

Assessing the Air Humidity Characteristics of Local Climate Zones in Guangzhou, China

Xiao Tan [1], Qi Zhang [1], Yiqi Chen [1], Junsong Wang [2], Lihua Zhao [2,3] and Guang Chen [1,*]

[1] School of Architecture and Urban Planning, Guangdong University of Technology, Guangzhou 510090, China; 17620079846@163.com (X.T.)
[2] School of Architecture, South China University of Technology, Guangzhou 510640, China
[3] State Key Laboratory of Subtropical Building and Urban Science, Guangzhou 510640, China
* Correspondence: urbanchg@gdut.edu.cn

Abstract: An urban canopy's humidity significantly affects thermal comfort, public health, and building energy efficiency; however, it remains insufficiently understood. This study employed 3-year (2020–2022) fixed measurements from Guangzhou to investigate the temporal patterns of relative humidity (RH), vapor pressure (Ea), and vapor pressure deficit (VPD) across eight local climatic zones (LCZs). Clear and distinct patterns in the humidity characteristics among the LCZs were revealed on multiple timescales. The *RH* and *VPD* of each zone were higher in summer than in winter, with peak RH observed in LCZ A (83.45%) and peak *VPD* in LCZ 3 (13.6 hPa). Furthermore, a significant daytime urban dry island (UDI) effect in the summer and a nighttime urban moisture island (UMI) effect in the winter were observed in terms of the Ea difference between urban and rural areas. The strongest UMI occurred during winter nights in LCZ 8, with a peak intensity of 0.8 hPa, while the UDI was more frequent during summer days in LCZ 1, with a maximum value of −1.2 hPa; meanwhile, compact areas had a slightly higher frequency of UDI than open areas. Finally, the effects of the urban heat island (UHI) and wind speed (*V*) on UMI were analyzed. During the daytime, a weak correlation was observed between the UHI and UMI. Wind enhanced the intensity of the nighttime UMI. This research offers further insights into the canopy humidity characteristics in low-latitude subtropical cities, thereby contributing to the establishment of a universal model to quantify the differences in moisture between urban and rural areas.

Keywords: urban moisture island; canopy humidity; local climate zone; fixed measurement; temporal characteristic

1. Introduction

Urbanization transforms the underlying surface of a city, replacing natural landscapes with impervious surfaces, resulting in changes in the heat and humidity environments of urban areas [1,2]. These changes include a decline in the latent heat flux [3], increases in the sensible heat flux and surface temperature, and a decrease in humidity [4]. The urban heat island (UHI) effect and urban moisture island effect (UMI) are urban microclimate modification phenomena [5]. Investigating urban climate issues, particularly in humid environments, is vital for improving living conditions, establishing a scientific evaluation system for the urban physical environment [6], and aiding urban planning and energy-efficient construction [7,8].

Traditionally, the UHI and UMI have been described through air temperature and humidity differences between urban and rural areas. However, this binary classification often

 https://doi.org/10.3390/buildings15010095

fails to capture intricate intra-urban and intra-rural temperature variations. To address this limitation, Stewart and Oke introduced local climate zones (LCZs) as a universal framework for urban climate research, providing a climate-based and objective classification of urban and rural landscapes [9]. This framework categorizes urban spaces into 17 distinct classes based on surface cover and structure (e.g., building height and spacing), facilitating nuanced comparisons and a deeper understanding of urban climate complexities.

Various methods are used to measure the impact of urbanization on changes in humid environments. The most widely used method is based on fixed observations at urban and rural meteorological stations, such as those in Beijing [10], Szeged [11], and Turin [12]. This method utilizes observational data from both urban and rural meteorological stations and can intuitively compare and analyze differences in humidity between urban and rural areas. While this method allows for straightforward comparison of urban and rural humidity, it is limited by environmental factors and the availability of meteorological stations. To better describe the spatial patterns of urban humidity, some researchers have employed mobile measurements to build urban humidity databases, such as by driving on specific routes in cities using automobiles fitted with meteorological sensors [13,14]. This method is cost-effective and provides spatial insights, but it cannot effectively capture the temporal dynamics of humidity [15]. Modeling is an effective way to study urban climate change by quantifying the factors influencing UMI. The urban canopy model (UCM) is often combined with weather research and forecasting (WRF) models to study the influence of the underlying urban surface and canopy on atmospheric motion. The impacts of street and canyon geometry, building emissions, and vegetation on outdoor air temperature (T), humidity, and thermal stress in Hong Kong were studied during a typical heat wave period based on the UCM, and it was found that as the street canyon height/width ratio increased, the daytime UHI phenomenon transformed into an urban cool island (UCI) phenomenon [16].

The observed humidity in urban areas may be influenced by several factors at different scales, such as albedo, the impervious plan area ratio, and average building height (HRE). The difference in humidity parameters in the same local area may be greater than that between regions [11]. In Nanjing, Yang, et al. [17] studied the spatial and temporal characteristics of the UMI in local climate zones (LCZs) based on measured data using the relative humidity (RH) and humidity ratio (q) as variables. The authors found that the difference in RH between urban and rural areas increased during strong nighttime UHI effects. The UDI was observed more frequently during the daytime in warm months. Based on the LCZ theory, Dunjic, et al. [18] studied the temporal characteristics of local-scale humidity (RH, vapor pressure [Ea], absolute humidity [AH], q, and vapor pressure deficit [VPD]) and the influence of T in Novi Sad City with fixed observation data from weather stations and found that the UDI effect frequently occurred in LCZ 2 from February to October. In addition, the UMI intensity reached its maximum during the daytime and at midnight during the heat wave. In Szeged City, Unger, Skarbit and Gál [11] calculated and compared the differences in Ea and temporal characteristics among different LCZs using data obtained from a city meteorological network based on the LCZ. The peak Ea occurred in the summer and the minimum value occurred in the winter. In Hong Kong, Du, et al. [19] used WUDAPT to obtain the urban morphology and land use and land cover (LULC) data to create an LCZ map, which was integrated into the WRF model to reveal the water transportation mechanism within the city. Weakened upward motion, blocked ventilation, and inhibited dew formation were the reasons for the higher moisture content in the cities than those in the suburbs at night. A positive correlation between the nocturnal UHI and UMI has been reported in Guangzhou [20], where the UMI intensity (UMII) decreased at sunrise and increased at sunset. Linear regression analysis showed that the UMII was proportional to the impervious plan area ratio and inversely proportional to the green cover ratio.

Despite these contributions, there are still deficiencies in understanding the complex dynamics of urban humidity, especially at the finer scales within individual urban zones. Most research has focused on thermal or general moisture patterns, yet few studies have comprehensively examined how multiple humidity parameters, such as RH, Ea, and VPD, interact in specific urban contexts. Additionally, temporal variations in canopy humidity across different LCZs have not been systematically studied. While LCZs offer a valuable framework for analyzing urban climates, understanding urban humidity dynamics requires a broader sample of cities, as the background climate of each city significantly influences its microclimatic patterns.

Establishing a robust database of urban humidity characteristics across various cities is crucial for refining LCZ theory and better understanding the mechanisms driving urban humidity changes. This study focused on Guangzhou, a representative humid subtropical city in southern China. Based on the LCZ theory, we selected eight representative zones and conducted long-term measurements of T and RH in each LCZ. We used RH, Ea, and VPD as variables and analyzed the temporal characteristics under different urban morphologies using 3 y of hourly observation data. This study contributes to establishing a scientific evaluation system for thermal and humid environments, which is of great significance in assisting urban planning and building energy conservation.

2. Methods

2.1. Study Area

Guangzhou, the provincial capital of Guangdong, is located in the south of China (23°08′ N, 113°15′ E) and is situated on the northern edge of the Pearl River Delta, facing the South China Sea. The city terrain is mountainous in the northeast, hilly in the central region, and comprises plains in the south. With an area of approximately 7434 km^2, Guangzhou has a population of 18.81 million residents, according to the 2021 census. According to the Köppen-Geiger classification system, Guangzhou has a humid subtropical climate (Cfa) characterized by hot and humid summers and cool and dry winters [20]. Based on the dividing region for building thermal design, Guangzhou is a hot summer and warm winter area, where the hottest month is July, with a mean T of 28.7 °C, and the coldest month is January, with a mean T of 13.3 °C. The dominant wind direction is southeast in the summer and north in the winter. Guangzhou has an annual average rainfall of approximately 1600–1900 mm and annual sunshine duration of 1500–2000 h. The total solar radiation during the summer is approximately 7318 w/m^2 (https://data.cma.cn/, accessed on 23 June 2023).

Guangzhou's urbanization, marked by changes in land cover and the expansion of built-up areas, has significantly altered local climate conditions. Meteorological data and prior research show that this rapid urbanization has contributed to a notable increase in the city's average temperatures [21]. Specifically, Guangzhou's annual average temperature has steadily increased since 1953, with rates accelerating from 0.22 °C per decade (1953–2009) to 0.38 °C per decade (1973–2009), and further to 0.49 °C per decade (1983–2009) [22]. This consistent and speeding up warming pattern indicates both the increasing intensity of the UHI and a clear trend of regional climate warming in Guangzhou. This warming trend, in turn, interacts closely with the city's diverse urban forms. From densely built-up high-rise areas to more suburban and green spaces, these various urban landscapes result in distinct microclimates that are strongly influenced by urbanization. The LCZ framework can effectively capture these variations, distinguishing zones with differing built environment and land cover characteristics and enabling a detailed analysis of urbanization's influence on canopy humidity.

2.2. LCZ Sites

To conduct an objective and standardized study of urban climates, Stewart and Oke [9] proposed 17 standard classes of local climates. LCZs are defined as regions with similar surface covers and surface structures (e.g., building height and spacing), with areas ranging from hundreds of square meters to several kilometers. This method helps to identify potential local climate problems and impacts caused by urban development.

We conducted LCZ zoning in Guangzhou using the method proposed previously [9]. First, by referring to the LCZ map of Guangzhou [23], satellite images, remote sensing data, and other information, we screened LCZs with similar surface structures and land cover. Each 500 m radius LCZ was selected to fully consider the influences of buildings, roads, and trees. Finally, on-site inspections and measurements of urban morphology parameters, such as HRE, sky view factors (SVFs), and aspect ratio (H/W), were performed. It was confirmed that the selected LCZ was in accordance with the identified LCZ type. GIS was employed to extract the land use and land cover types for selected regions within Guangzhou, and the specific locations of each zones were subsequently identified, as seen in Figure 1.

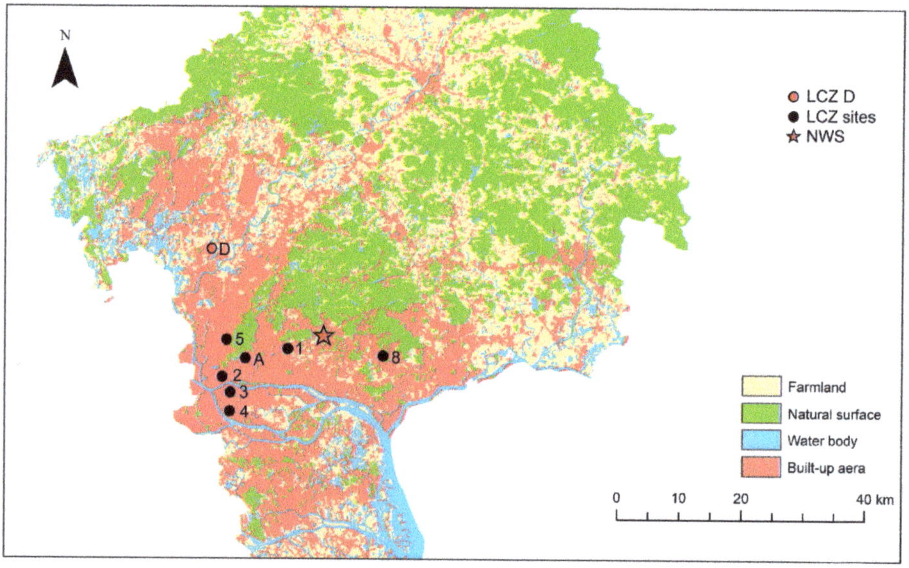

Figure 1. Land use/land cover (LULC) and the distribution of each LCZ site in Guangzhou. The dots represent the locations and types of each LCZ, while the star indicates the location of national meteorological station (NWS).

The HRE, SVF, H/W, building density (BSF), and impervious surface area ratio (ISF) of the LCZ morphological indicators are listed in Table 1. Specific methods for calculating the morphological indicators of each LCZ can be found in a previous study [24]. It can be observed that the morphological parameters of the eight zones generally fall within the defined range, and the locations of these districts are also consistent with the classification results obtained by WUDAPT [25].

Table 1. Types and morphological indicators of eight local climate zones (LCZs).

LCZ Type	Urban Morphological Indicator and Their Range of LCZ Categories				
	H/W	**SVF**	**BSF (%)**	**ISF (%)**	**HRE (m)**
1 Compact high-rise	2.24 (>2)	0.48 (0.2–0.4)	30.55 (40–60)	49.28 (40–60)	35 (>25)
2 Compact mid-rise	1.06 (0.75–2)	0.526 (0.3–0.6)	42.70 (40–70)	54.69 (30–50)	19 (10–25)
3 Compact low-rise	0.84 (0.75–1.5)	0.57 (0.2–0.6)	44.86 (40–70)	51.61 (20–50)	11 (3–10)
4 Open high-rise	1.13 (0.7–1.25)	0.56 (0.5–0.7)	26.14 (20–40)	56.93 (30–40)	28 (>25)
5 Open mid-rise	0.83 (0.3–0.75)	0.51 (0.5–0.8)	28.21 (20–40)	55.27 (30–50)	20 (10–25)
8 Large low-rise	0.18 (0.1–0.3)	0.73 (>0.7)	36.31 (30–50)	35.60 (40–50)	5 (3–10)
A Dense trees	2 (>1)	-	4.4 (<10)	1 (<10)	13 (3–30)
D Low plants	<0.1 (<0.1)	0.9 (>0.9)	4.8 (<10)	2.7 (<10)	<1 (<1)

H/W: aspect ratio; HRE: average building height; SVF: sky view factor; BSF: building density; ISF: impervious surface area ratio.

2.3. Data

The monitoring experiment was conducted continuously from 1 January 2020, until 31 December 2022. Two temperature–humidity data loggers (HOBO MX2301, Onset Computer Corporation, USA), each with accuracies of ±0.2 °C and ±2.5%, were installed in each LCZ plot, and the sampling rate was set to once per hour. To obtain accurate and representative data, both instruments were placed within a 100 m radius of the LCZ center and 100 m apart, and hourly data were collected. The drift of the loggers was found to be less than 0.01 °C and 0.01% for RH, which is considered negligible for the purpose of this study. To further ensure the reliability of the data from each zone, the average value of the readings from these two loggers is then taken as the basis for subsequent analysis. These loggers ensured measurement accuracy in harsh outdoor environments. Each HOBO data logger was equipped with a radiation shield to minimize the effect of radiation during the measurement process. The installation positions of the loggers were well-ventilated, away from heat sources, and avoided being blocked by surrounding obstacles, such as buildings and trees. These instruments were installed on streetlights or utility poles at a height of 2.5 m, as far from vehicles and air conditioners as possible, and at least 3 m from the walls.

LCZ D was used as a reference site for rural areas. It is located far from the city center and is mainly covered by low plants and farmland. Hence, it is representative of typical non-urbanized landscapes [17]. Figure 2b presents one of the measurement points (D2) situated within LCZ D, with the photograph captured during the on-site instrumentation setup.

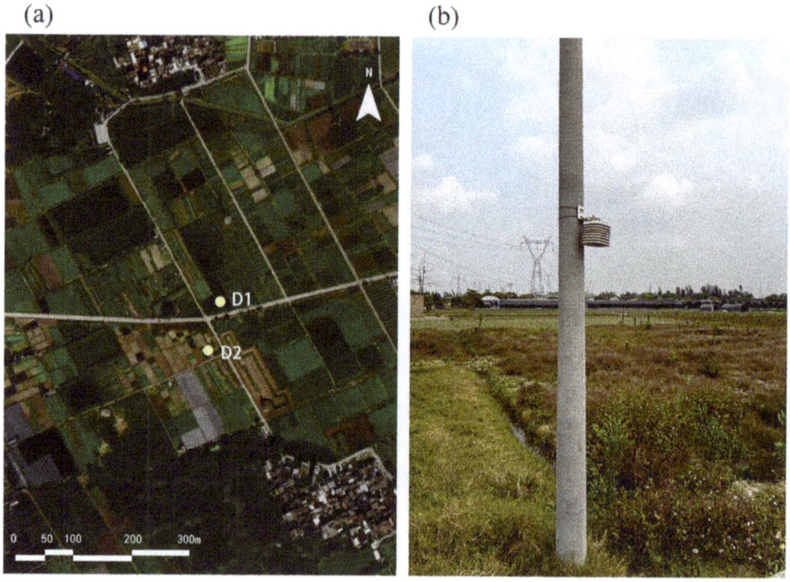

Figure 2. (**a**) Satellite imagery of local climate zone D (LCZ D, characterized by low plants) sourced from Google Maps (https://www.amap.com, accessed on 26 December 2024); "D1" and "D2" represent the detailed locations of the two loggers. (**b**) The specific locations where the temperature and humidity data loggers were installed.

2.4. Typical Days for UMI Development

The dynamic and magnitude of humidity differences between urban and rural areas exhibit similarities to the UHI. The maximum intensity of both phenomena is observed on windless, clear summer nights, indicating that the UMI develops most favorably under "ideal" weather conditions [24]. It is important to investigate the temporal humidity patterns on such days to study UMI formation and development. Oke proposed weather factors (a calculation formula based on wind speed, cloud cover, and cloud shape) to identify typical meteorological days [24]; this method has been effectively utilized in studies on thermal behavior and humidity characteristics in Nanjing [17,26]. However, acquiring certain meteorological parameters using this method is challenging. Our previous work validated the significant linear correlation between the rural diurnal temperature range (DTR) and the daily maximum urban heat island intensity, with a Pearson correlation coefficient exceeding 0.7 [27]. A threshold for identifying typical meteorological days, DTR $\geq$ 10 °C, has been established [27]. Specifically, DTR is measured by subtracting the hourly minimum temperature from the maximum hourly temperature in a day (LCZ D). It serves as an indicator of the combined influence of various meteorological factors in rural areas and can be utilized to assess variations in the local thermal and humid environment. Therefore, in the present study, the DTR of the LCZ D was used to identify typical meteorological days.

Specifically, days with accumulated precipitation of >0.1 mm were excluded, and hourly precipitation data during the study period were downloaded from the Guangzhou National Meteorological Station (No. 59287). Days with fog (daily mean RH $\geq$ 80%) and days with hourly variations in wind speed exceeding 2 m/s were excluded [28]. Finally, 275 days with DTR $\geq$ 10 °C were selected as typical meteorological days for this study. The numbers of such days in each year and month are presented in Table 2.

Table 2. Number of typical meteorological days during the study period (2020.1.1–2022.12.31).

	Jan	Feb	Mar	Apr	May	Jun	Jul	Aug	Sep	Oct	Nov	Dec	3 y Total
2020	0	6	2	9	3	3	7	4	1	5	8	11	
2021	20	16	13	11	5	5	3	3	9	6	16	14	275
2022	11	0	10	11	5	2	10	8	19	13	1	5	

To study the differences in the UMI between the day and night, 6 a.m. to 6 p.m. was designated as daytime and 6 p.m. to 6 a.m. on the following day was designated as nighttime.

2.5. Humidity Parameters

Understanding the distribution and variation of urban humidity requires analysis of both the relative moisture content (in relation to the current atmospheric state) and the absolute moisture content (under standard or typical conditions). Although there is no standard for selecting specific indices, most studies use different indicators to capture these two aspects of urban humidity. This study employed RH, Ea, and VPD as key indicators. These indices provide complementary perspectives by addressing both the absolute characteristics of humidity (e.g., UMI or UDI) and the relative characteristics (for comparative analysis), which ensured the comparability and representativeness of the research conclusions.

- RH

RH is the most commonly used parameter for measuring humidity owing to its ease of measurement and comparability. RH significantly impacts urban rainfall, evaporation, and human thermal comfort. It represents the percentage of vapor pressure in the air relative to the saturated vapor pressure (Es) at the same temperature and atmospheric pressure. This information can be obtained directly using a HOBO data logger.

- Ea

The actual water content in air can be assessed using various humidity metrics. Ea directly reflects the vapor content in the air, which does not change with variations in air temperature. However, the difference in temperature can affect evaporation and condensation, and can be reflected by changes in indicators, such as Ea. Ea can be calculated using Es, as follows:

$$\mathrm{Ea} = Es \times \frac{RH}{100} \quad (1)$$

- VPD

VPD is an indicator that measures the difference between actual and saturated vapor in air. It represents air dryness and is an important indicator for studying water circulation in urban climates. Similar to the RH, VPD is sensitive to temperature. During the UHI period, higher temperatures can cause an increase in the Es, leading to a water vapor shortage. VPD can be calculated by subtracting Ea from Es, which can be obtained using the Bolton equation [29]. The Equation (3) was adjusted to match Wexler's results and was extrapolated for temperatures less than 0 °C. It has an accuracy of 0.1% within the temperature range from −30 °C to 35 °C [30].

$$VPD = \mathrm{Es} - Ea \quad (2)$$

$$Es = 6.112 \times \exp\left(\frac{17.67 \times T}{243.5 + T}\right) \quad (3)$$

- Definition of UMI

The UMII is usually calculated using Ea, RH, AH, and the humidity ratio. In this study, vapor pressure was selected as the only indicator to measure the UMII, i.e., the difference between the Ea of each LCZ and that of the corresponding LCZ D (ΔEa and hPa). If the value was positive, it was a UMI; otherwise, it was a UDI. Meanwhile, ΔRH and ΔEa are only used as variables to measure humidity characteristics.

$$\Delta \mathrm{Ea} = Eurban - Erural \tag{4}$$

3. Results

3.1. Temporal Pattern of Humidity

Figure 3 depicts the characteristics of RH, Ea, and VPD for each LCZ in terms of seasonal, and diurnal features. The differences in RH among the zones were not significant; however, the seasonal and diurnal variations showed higher humidity in the summer than in the winter, and at night than during the day. During seasonal variations, higher Ea was observed in all zones in summer, while the differences between LCZs within the same season were not significant. Specifically, the average distribution of Ea across all LCZs in summer exhibited a "slight" diurnal variation, with higher values during the day and lower values at night, but the change did not exceed 5 hPa for both daytime and nighttime. It can be observed that the RH, Ea, and VPD in all zones are higher in summer than in winter. This is reasonable and can be explained as follows: (1) Summer is the season with relatively concentrated precipitation in Guangzhou. Although the Es increases with temperature, the frequent precipitation caused by the influence of warm and humid ocean currents is the dominant factor for the increase in RH and Ea. (2) In summer, the high temperature causes intense evaporation of water in surface water bodies and soil. This process is significantly weakened in winter. (3) Due to the higher Es in summer than in winter, even if the water vapor in the air is closer to saturation, the "distance" from saturation can still be higher than in winter. During the summer and summer nights, each LCZ showed a higher average VPD than in the winter and summer days, indicating that the air was drier during these periods. In the winter, the difference in air dryness between the LCZs was not significant.

It can be concluded that the RH in Guangzhou was higher during the summer and at night throughout the year, and that Ea was significantly higher in the summer than in the winter. However, there was minimal difference in the daily variation.

3.2. Temporal UMI Pattern

3.2.1. Monthly Variations

The humidity differences between each urban LCZ and LCZ D were analyzed on a monthly basis (LCZ X—LCZ D). Figure 4 shows the monthly mean ΔRHX-D and ΔEaX-D during the daytime and nighttime. To further explore the statistical significance of these differences, an ANOVA was performed on the ΔRHX-D and ΔEaX-D across the six LCZs. Table 3 presents the results of this ANOVA analysis, summarizing the statistical differences for each time period and LCZ zone, with F-values, degrees of freedom, and *p*-values provided for each comparison.

The results showed that the ΔRHX-D was negative for most of the year (except for LCZ 8), which indicated that the urban environment was perceived as drier than that in rural areas. This dryness was more pronounced in the summer and at night. The smallest ΔRHX-D occurred at nighttime in September (−8.7% in LCZ 3). In the comparison between urban and rural areas, for each zone except LCZ 8, the RH is lower in every period throughout the year. This trend reaches its peak in summer. The high RH in LCZ 8 throughout the year may have been due to its proximity to the water body, as the river was located within the radius of the two stations. Continuous evaporation would not only reduce the local temperature, but also increase the Ea. Although the perceived humidity in urban areas is lower than that in rural areas, in terms of the actual water vapor content, this is not the case.

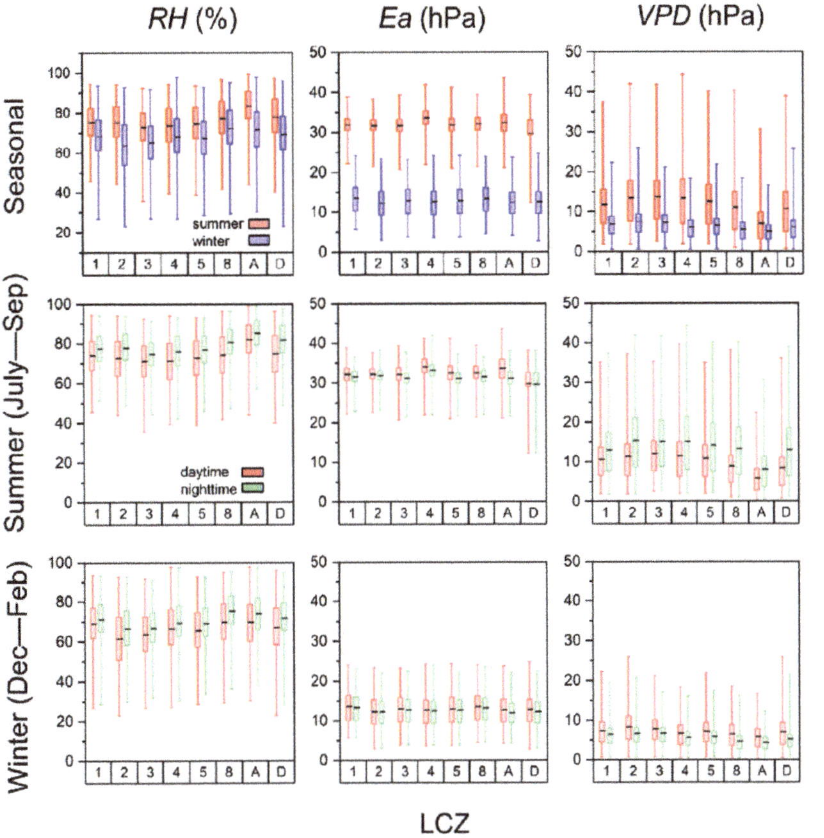

Figure 3. Distributions of RH, Ea, and VPD for LCZ 1–LCZ D during the season, and diurnal cycles. All the sample of three years of typical days were used for analysis (n = 275 days).

The ΔEa pattern exhibited a more remarkable difference throughout the year than ΔRH. Most LCZs exhibited a stronger UMI on winter nights than on summer nights. The UMI gradually weakened with increasing temperature and eventually transformed into a UDI in August and September. Daytime data throughout the year showed that the UDI dominated during warm months. From the perspective of monthly averages, most urban areas exhibit a UMI throughout the day from January to April. In the hotter month of August, a UDI is presented throughout the day. For most months of the year, a daytime UDI and nighttime UMI are manifested.

The impervious surface area in urban environments, such as roads, buildings, and pavements, plays a significant role in the UDI effect. These surfaces absorb and store heat during the day, leading to higher temperatures and increased evaporation. This results in lower vapor pressure in urban areas compared to rural zones, contributing to the UDI effect, especially during the day. This is particularly evident in LCZ 4, where due to its largest ISF among the six zones (Table 1), a UDI occurs throughout the day throughout the year.

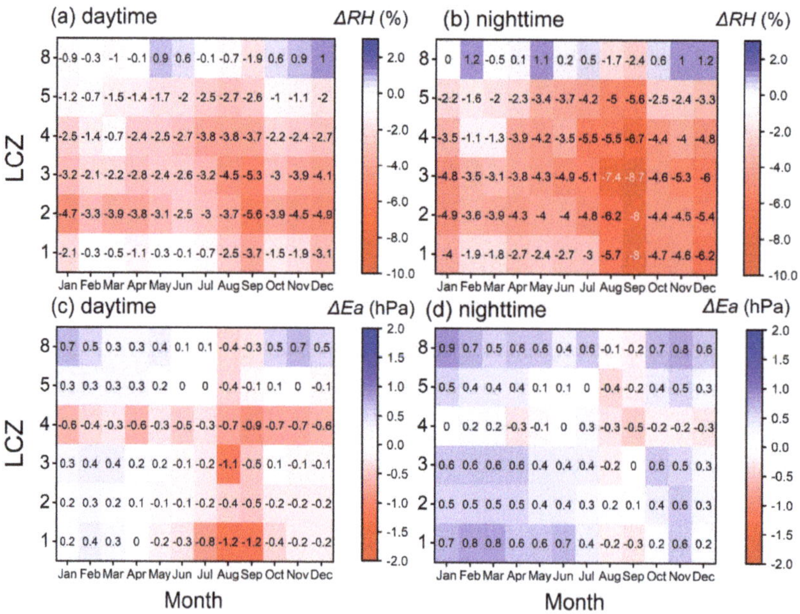

Figure 4. Monthly urban—rural differences in RH in the (**a**) daytime and (**b**) nighttime and in ΔEa in the (**c**) daytime and (**d**) nighttime under all weather conditions (n = 875 days).

Table 3. ANOVA Analysis of Humidity Parameters in LCZs.

LCZ Type	ANOVA Results				
	Source	Sum_sq	df	F	PR (>F)
Daytime ΔRH	LCZ	9.38774×10^4	5.0	710.2	0.0
	Residual	1.48106×10^6	56,023.0	-	-
Nighttime ΔRH	LCZ	1.07122×10^5	5.0	810.6	0.0
	Residual	1.05777×10^6	40,020	-	-
Daytime ΔEa	LCZ	626.441951	5.0	131.3	8.9519×10^{-139}
	Residual	53,471.38311	56,023	-	-
Nighttime ΔEa	LCZ	785.23118	5.0	288.2	5.6970×10^{-304}
	Residual	21,809.35418	40,020	-	-

3.2.2. Hourly Variation

To further investigate the local-scale differences in UMI, Figure 5 presents the all-day variations in ΔRHX-D and ΔEaX-D for each LCZ during the summer (Figure 5a,c) and winter (Figure 5b,d). ΔRH and ΔEa were defined as the differences in relative humidity and vapor pressure, respectively, between urban and rural areas. The diurnal variations in ΔRH were similar across all LCZs, with the exception of LCZ 8; thus, the humidity levels of urban areas can be represented by the characteristics of most LCZs. In winter, the humidity characteristics of urban areas demonstrate the following pattern: RH is lower than that of rural areas from 00:00 to 12:00, higher from 12:00 to 18:00, and returns to a drier state after 18:00. This suggests that urban areas experience relatively higher humidity during specific daytime periods, although this condition is not consistently maintained. Throughout the day, ΔRH decreases gradually in the initial 2 h following sunrise (6:00 in the summer and 8:00 in the winter) and increases during the pre-sunset period (15:00 in the summer and the winter). Except for the significant temperature fluctuations during sunrise and sunset, the disparities in RH between the urban and rural areas remained stable and were mostly maintained at levels within a range of 0% to −10% during the night (excluding LCZ 8). This trend was observed for both summer and winter, with no distinguishable differences.

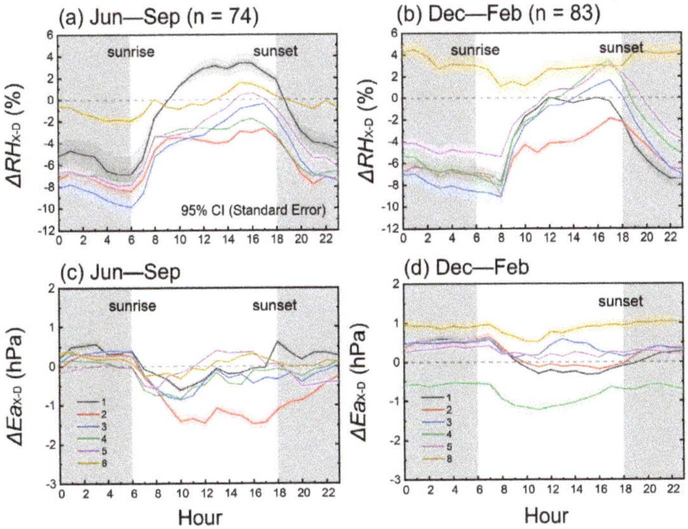

Figure 5. Daily variations in urban—rural differences in relative humidity (ΔRHX-D) and vapor pressure (ΔEaX-D) for typical meteorological days in (**a**,**c**) summer (June–September); (**b**,**d**) winter (December–February).

Based on the ΔEa curve, all LCZs in the summer exhibited nighttime UMI and daytime UDI. Compared with winter, a larger diurnal variation in UMII (up to −2.4 hPa to 1.28 hPa) was observed for each LCZ in the summer. During the first two hours after sunrise, a trend of initial decrease followed by an increase was observed in ΔEa across all zones. In simpler terms, this indicates a weakening of the UMI and a strengthening of the UDI. The decrease in UMII observed during sunrise could be attributed to evapotranspiration from residual dew and vast vegetation in rural areas. The evaporation process continued for 2 h after sunrise; however, the increase in the mixing layer height and strengthened upward motion resulted in a decrease in the observed Ea. The higher SVF in rural areas and increased surface cooling led to greater condensation and a continuous increase in ΔEa after sunset. Overall, whether in summer or winter, the state of higher RH in urban areas is transient. However, from the perspective of Ea, urban areas consistently exhibit greater moisture levels than rural areas throughout the winter.

3.3. UMI Frequency

According to the definition of Kuttler et al., 0 hPa < ΔEa < 0.5 hPa is defined as a weak UMI, and ΔEa > 0.5 hPa is defined as a strong UMI [31]. The frequency characteristics of three types of ΔEa (weak UMI, strong UMI, and UDI) and two types of ΔRH (ΔRH < 0 and ΔRH > 0) in the four seasons were analyzed (Figure 6). Here, a state with UMII > 0 is defined as urban moisture excess (UME).

Figure 6 shows the statistics of ΔRH and ΔEa (positive or negative) in the four seasons. The types of ΔRH did not differ significantly among the four seasons, with ΔRH > 0 accounting for an average of 26% and ΔRH < 0 accounting for an average of 74% in one season. As the ΔRH trend in a day was basically consistent (Figure 5), the seasonal pattern of RH differences between urban and rural areas was mainly reflected in the magnitude of the value, rather than the type (positive or negative). Notably, LCZ 8 consistently exhibited

ΔRH values greater than zero throughout winter (Figure 5). Despite the fact that the presence of such high ΔRH values may elevate the overall average or alter the distribution of results, the impact of larger building surface areas and smaller vertical heights on the distribution of urban moisture should not be overlooked. Low-rise buildings typically exert a significant influence on wind, restricting vertical air exchange and promoting the accumulation of moisture near the ground. The ΔEa curve showed obvious differences among the four seasons. The UMEs were dominant throughout the four seasons. The frequency of strong UMIs was the highest from March to May (67%). A relatively dry period occurred from October to November, with a UDI frequency of 37% (Figure 6b).

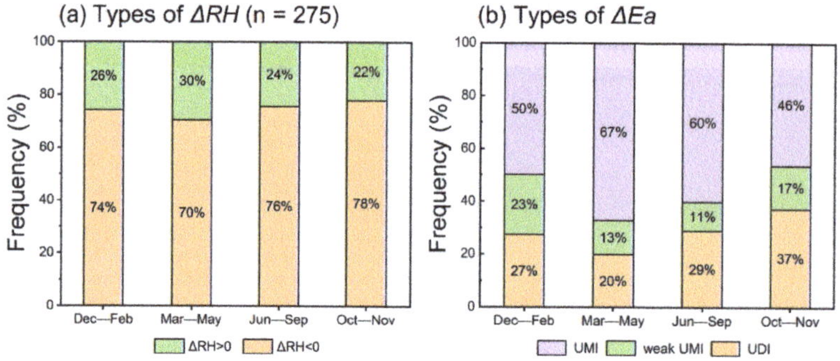

Figure 6. Frequencies of different types of ΔRH (ΔRH < 0% and ΔRH > 0%) and ΔEa (ΔEa < 0 hPa, ΔEa < 0.5 hPa, ΔEa > 0.5 hPa) in all LCZs on typical meteorological days during the four seasons. (**a**) Frequencies of ΔRH types; (**b**) Frequencies of ΔEa types.

Figure 7 shows the proportion of ΔEa types for each LCZ in the summer (Figure 7a) and winter (Figure 7b). It can be clearly observed that the UDI frequency in most LCZs (except for LCZ 4) was higher in the summer, and the frequency of weak UMI was higher in the winter. The UDI frequency decreased gradually with a decrease in the HRE of the LCZ, and compact areas had a slightly higher UDI frequency than open areas. This trend was more pronounced during the summer. The dense high-rises blocked airflow, causing heat and moisture to diffuse slowly, resulting in the formation of a high-local-RH environment, an increase in the latent heat, and a decrease in Es. This decreased the amount of moisture contained in the air. However, as the HRE decreased (LCZ 2 and LCZ 3), the effect gradually weakened.

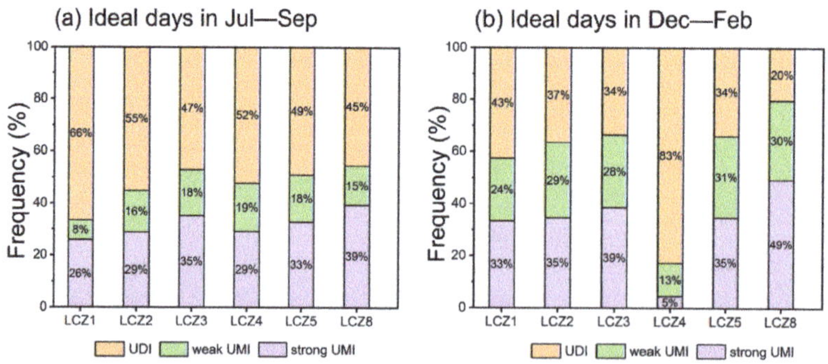

Figure 7. Frequencies of urban dry island (ΔEa < 0 hPa), weak urban moisture island (0 hPa ≤ ΔEa < 0.5 hPa), and strong urban moisture island (ΔEa ≥ 0.5 hPa) in each LCZ on typical meteorological days during (**a**) summer and (**b**) winter.

3.4. *Influencing Factors of UMI*

3.4.1. UHI

To investigate the relationship between the UHII and UMII, data from LCZ 1 and LCZ D were used for analysis. Due to partial data loss during the measurement process of LCZ 1, analysis was conducted utilizing the remaining 120 typical days. Seasonal and diurnal conditions were investigated, including those in the summer daytime (Figure 8a), summer nighttime (Figure 8b), winter daytime (Figure 8c), and winter nighttime (Figure 8d).

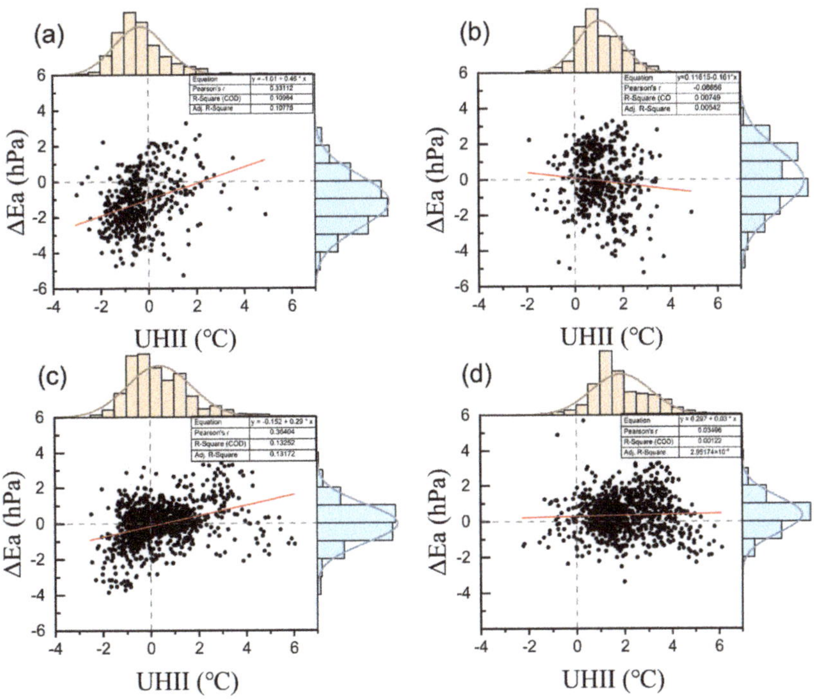

Figure 8. Relationships between UHII and urban—rural vapor pressure difference (ΔEa) in LCZ 1 in (**a**) summer daytime, (**b**) summer nighttime, (**c**) winter daytime, and (**d**) winter nighttime during typical days (120 d in total).

During the daytime in the summer (Figure 8a), the majority of the scatter was distributed in the lower-left quadrant of the plot (ΔEa < 0 and UHII < 0). A weak correlation between UHI and UMI can be found in daytime in summer (Figure 8a) and winter (Figure 8c), which was weaker during nighttime. This correlation was not evident in the comparison of different seasons. During the daytime, a decrease in the intensity of the UHI implied that evapotranspiration in the suburbs was stronger than that in urban areas, which directly led to an decrease in the UMII.

3.4.2. Wind Speed

Figure 9 provides statistics on V and the corresponding ΔEa of all LCZs on typical days, with the aim of investigating the influence of wind speed magnitude on the UMI. To enhance representativeness, wind speeds from 0 to 6 m/s were selected and divided into datasets with 1 m/s increments. The ΔEa corresponding to the V at each moment is categorized and averaged under different wind speed classes. Daytime (Figure 9a) and nighttime (Figure 9b) conditions are discussed.

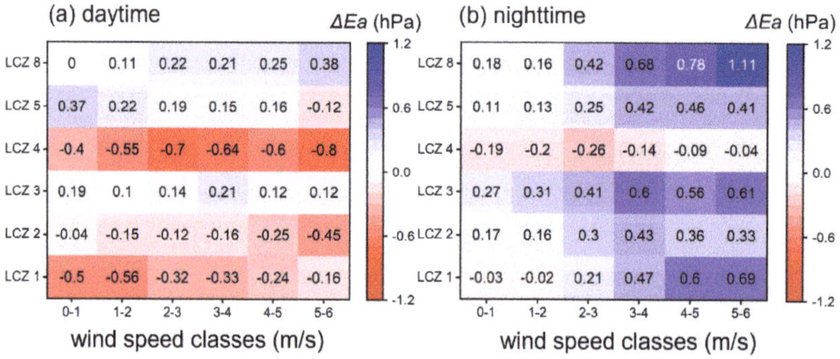

Figure 9. Relationship between wind speed and urban—rural vapor pressure difference (ΔEa) in all LCZs (LCZ 1—LCZ 8) during typical weather days in (**a**) daytime and (**b**) nighttime.

During the daytime, the relationship between ΔEa and V was not clearly defined. In some LCZs (e.g., LCZ 2, 4, 5), ΔEa decreased with increasing V, while in others (e.g., LCZ 1, 8), it increased. This randomness can be attributed to the varying influence of V on local humidity in different LCZs, which may be affected by the surrounding areas. While the air flow does influence temperature and evapotranspiration, thereby altering humidity levels, the underlying mechanisms remain unclear and require further data to support a more robust understanding.

At night, a more consistent positive correlation between ΔEa and V was observed in LCZs 1, 2, 3, 5, and 8, where ΔEa increased with wind speed. However, this trend exhibited changes at a "threshold", such as a decrease in LCZs 2 and 5, and fluctuation in LCZ 3. The emergence of this threshold effect can be attributed to the fact that excessively high wind speeds result in a more pronounced dissipation to evapotranspiration.

LCZ 1 typically exhibits a higher threshold compared to LCZs 2, 3, and 5 due to its increased surface roughness and more localized airflow dynamics. This leads to the creation of more turbulence and enhanced moisture exchange at higher wind speeds, thereby allowing for the UMI to persist for a longer duration. In contrast, an explanation for the lower threshold in LCZs 2, 3, and 5 could be related to the role of evaporative cooling and the influence of UHI. In LCZs with more open or mixed building structures, there is often greater exposure to solar radiation and urban heat island effects, which leads to higher surface temperatures and greater evapotranspiration during the day. As a consequence, when wind speeds increase, they can disrupt the localized moisture accumulation faster, causing a more immediate response in humidity changes. In contrast, in denser LCZs like LCZ 1, where buildings are closer together and shaded areas are more common, there is more localized heat retention, and the wind has to be stronger before it can disrupt the more stable moisture conditions in place.

4. Discussion

4.1. Reference Rural Site

Rural areas often exhibit lower T and higher RH than urban areas due to differences in multiple surface characteristics, such as radiation (albedo and radiance), LULC, and aerodynamics (roughness) in rural areas [32,33]. It should be noted that the peak Ea usually appears several hours before and after sunrise and sunset in rural areas, as reported in many studies [11,12,17]. The appearance of these "two peaks" is related to the contraction and expansion of the urban boundary layer during sunset and sunrise, respectively. In this study, the same phenomenon was observed in the diurnal cycle of the LCZ D from June to September; however, the second peak was not pronounced.

The selection of rural reference sites directly affected the results and characteristics of the UMI. This also explains the diversity in UMIs in different cities, as the selection of reference rural sites varies across studies. In Nanjing [17], when using LCZ D as the reference rural site, a diurnal UDI and a nocturnal UMI were present in January, whereas the UDI was present throughout the day in July. In Hong Kong [34], two rural meteorological stations (one in a rural residential area and the other in a forested area) were used as reference sites. The nocturnal UMI performance was similar between the two reference sites; however, the daytime UMI was larger when the rural residential area was used as the reference site. In Novi Sad [18], LCZ A was used as the reference rural site because of its high evapotranspiration rate. The largest UDI intensity was observed in July, whereas a slight UME was observed in colder months, similar to Guangzhou.

In selecting a reference rural site, the most important aspect is to ensure that all sites are located under the same climatic conditions and terrain [35], and to preferably locate rural stations outside the range of urban climate impact [36]. If these conditions are met, the differences in atmospheric variables between urban and rural areas can be used to quantify urban climate effects [24].

4.2. Humidity-Based Indices

Studying the amount of air humidity and moisture transfer in urban environments must involve considerations of the saturation level and specific quantity of water vapor. This is because the urban vapor circulation process involves various physical processes, such as vegetation evapotranspiration [37], dew condensation [38], and precipitation [39], which cannot be completely measured by a single variable. For example, the rate of evapotranspiration is related to both RH and Ea.

Generally, the lower the RH, the faster the surface evaporation rate [40]. This is because a lower RH causes a more rapid diffusion rate of water molecules into the air and, correspondingly, a faster surface evaporation rate. However, the RH cannot directly reflect the surface evaporation rate on urban surfaces, because it only describes the content and degree of saturation of water molecules in the air. The surface evaporation rate is affected not only by RH, but also by meteorological factors, such as temperature, wind speed, and solar radiation [41]. Additionally, Ea is an important factor that affects the evaporation process. If Ea in the air is lower than Es on the water surface, the evaporation rate is relatively fast until Ea is equal to Es on the water surface, after which the evaporation rate slows down or stops. Overall, both RH and Ea affect the evaporation rate, and their combination can describe the evaporation process more accurately.

VPD describes the difference between Ea and Es, which better reflects the movement and transmission of water molecules [42]. In urban environments, dry and hot conditions can cause an increase in VPD, making it easier for water molecules to evaporate from vegetation and enter the atmosphere [43]. This leads to an increase in the evaporation rate of urban vegetation, which is particularly significant in the summer. However, for the same VPD, there may be two different Ea and Es values at different temperatures (i.e., VPD values may be the same when Ea and Es are both large or small). Therefore, by integrating RH and VPD, we can better describe the moisture content, thermal comfort, and other aspects, thus providing a more comprehensive assessment of humidity.

4.3. Humidity Characteristic

In extensive research on UMI and humidity variations, the role of urban morphology, green space coverage, and climatic conditions in influencing local humidity has been thoroughly examined. For instance, studies conducted in Beijing have shown that urban expansion leads to a decline in RH, Ea, and specific humidity within urban areas, particularly during the spring and autumn when dry conditions become more pronounced [44]. This trend is consistent with the seasonal variations in this study. Similar findings from Guangzhou and Novi Sad highlight the more pronounced UMI at night [18,20], a similar pattern also observed in this study, where the UMII is most prominent during the nighttime, particularly in high-density built-up zones such as LCZ 1 and LCZ 2.

Urban morphology features such as building density, green cover, and surface albedo have been shown to significantly influence the development of both UHI and UMI. For example, increasing green cover in LCZ 2 and LCZ 3 can notably enhance cooling and humidification effects [45]. Additionally, research on sky conditions has demonstrated that factors like solar radiation and cloud cover substantially impact local climate patterns [20]. On clear days, the UHI tends to intensify during the daytime, while overcast or rainy conditions promote moisture retention and latent heat release.

Table 4 summarizes the key findings related to UMI and humidity characteristics from the relevant studies. These findings underscore the complex interactions between urban morphology, climatic conditions, and the UMI. Building upon previous work, this research expands the understanding of canopy humidity dynamics in humid subtropical regions by exploring UMI patterns across various temporal scales and identifying key influencing factors. Based on these findings, future urban planning strategies should incorporate a holistic approach to mitigating both UHIs and UMIs. Specifically, urban designs should consider elements such as wind speed, urban morphology, and seasonal fluctuations to effectively reduce both heat and moisture imbalances and improve thermal comfort in urban areas.

Table 4. Summary of urban humidity researches.

Ref.	Location	Theme	Method
[12]	Turin	Temporal—spatial characteristics of UDIs and the impact of wind	Fixed measurement
[16]	Hong Kong	Mechanism of synergistic UHIs and UMIs	Numerical modeling
[18]	Novi Sad	Relationship between T and RH, AH, Ea, q, VPD, and the humidity characteristics under special weather conditions	Fixed measurement
[20]	Guangzhou	Characteristics and mechanisms of UMIs under different urban morphology and sky conditions	Fixed measurement
[34]	Hong Kong	Effects of urbanization (ventilation and land use) on humidity environments and temporal—spatial characteristics of UMIs	Numerical modeling
[44]	BTHUA	Long-term trend of atmospheric humidity and the urban expansion effect	Fixed measurement
[45]	Beijing	Evaluating the cooling and humidifying effects of urban vegetation coverage across different LCZs	Fixed measurement
[46]	China	Impact of urbanization on the humidity environment and mechanisms of climate change that affect UDIs and UMIs	Fixed measurement
[47]	Beijing	Seasonal and spatial characteristics of UMIs and UDIs, and their causes	Numerical modeling
[48]	15 cities	Differences in ΔT and Δq between the urban and rural sites (LCZ X − LCZ D) were analyzed on multiple time scales (diurnal, seasonal, and annual)	Fixed measurement
[49]	Barcelona	The impact of model resolution and urban land use information (LCZs) on humidity predictions	Numerical modeling

Note: T: temperature (°C); RH: relative humidity (%); AH: absolute humidity (g/m^3); Ea: vapor pressure (hPa); q: humidity ratio (g/kg); VPD: vapor pressure deficit (hPa).

5. Conclusions

This research utilized 3-year fixed measurements to study the characteristics of RH, Ea, and VPD across eight LCZs over varying time spans, with a particular focus on the temporal patterns of urban–rural differences, as well as the factors that influence UMIs. In most LCZs, RH, Ea, and VPD were higher in summer, with RH being higher during the nighttime, while the differences in Ea and VPD between the daytime and nighttime were not significant. From the perspective of urban–rural differences, the RH at urban stations was lower than that at rural stations every month, with this trend being most pronounced during summer nights. The UMI was more pronounced during winter nights and then transformed into a UDI during the summer. Regarding daily variation, ΔRHX-D increased after sunrise and decreased before sunset, with no notable seasonal features. In terms of the factors influencing the UMI, it was discovered that (a) a weak correlation between the UHI and UMI can be observed during the day in both winter and summer, and (b) nocturnal UMII increased with an increase in wind speed. The frequency analysis revealed that the highest frequency of strong UMIs occurred from March to May, whereas the highest frequency of UDIs occurred from October to November. Finally, the frequency of UDIs decreased with a decrease in the HRE, and compact areas had a higher frequency of UDIs than open areas. The identified characteristics of UMIs and UDIs cannot be universally classified as "good" or "bad". Instead, they should be considered in the context of urban planning aimed at creating comfortable spaces. Therefore, managing key factors influencing UMIs, such as the UHI and wind dynamics, is essential. For instance, by optimizing building layouts, urban planners can regulate wind distribution across seasons, ensuring that wind speed remains within optimal thresholds for each LCZ to control temperature and humidity, thereby enhancing thermal comfort.

Although this study analyzed the canopy humidity for each type of climate zone, the conclusions drawn may have certain limitations in terms of applicability. For example, the observed data obtained from different sites may vary, and the impact of natural and human factors on humidity can lead to errors and biases in the data. Based on the different humidity characteristics of the various LCZs in Guangzhou examined in this study, a more accurate humidity model could be established for further research. Building on the LCZ, more intricate and detailed models can be developed by further incorporating factors such as topography, vegetation, and land use, leading to improved accuracy in prediction and diagnosis.

Author Contributions: Conceptualization, X.T., Y.C. and G.C.; methodology, X.T., Q.Z., L.Z. and G.C.; software, X.T. and Q.Z.; validation, J.W., L.Z. and G.C.; formal analysis, X.T., Q.Z., Y.C. and G.C.; investigation, X.T., Q.Z. and G.C.; resources, J.W. and G.C.; data curation, X.T., Q.Z. and Y.C.; writing—original draft preparation, X.T., Q.Z., Y.C. and G.C.; writing—review and editing, X.T., Q.Z., Y.C. and G.C.; visualization, Q.Z. and Y.C.; supervision, G.C.; project administration, J.W., L.Z. and G.C.; funding acquisition, G.C. All authors have read and agreed to the published version of the manuscript.

Funding: This research was funded by National Key Research and Development Program of China, funding number 2022YFF1303105-4.

Data Availability Statement: Data are contained within this article.

Conflicts of Interest: The authors declare no conflicts of interest.

References

1. Qian, Y.; Chakraborty, T.C.; Li, J.; Li, D.; He, C.; Sarangi, C.; Chen, F.; Yang, X.; Leung, L.R. Urbanization Impact on Regional Climate and Extreme Weather: Current Understanding, Uncertainties, and Future Research Directions. *Adv. Atmos. Sci.* **2022**, *39*, 819–860. [CrossRef]
2. Ren, Z.; Fu, Y.; Dong, Y.; Zhang, P.; He, X. Rapid urbanization and climate change significantly contribute to worsening urban human thermal comfort: A national 183-city, 26-year study in China. *Urban CLim.* **2022**, *43*, 101154. [CrossRef]
3. Hou, H.; Liu, K.; Li, X.; Chen, S.; Wang, W.; Rong, K. Assessing the urban heat island variations and its influencing mechanism in metropolitan areas of Pearl River Delta, South China. *Phys. Chem. Earth* **2020**, *120*, 102953. [CrossRef]
4. Luan, Q.; Cao, Q.; Huang, L.; Liu, Y.; Wang, F. Identification of the Urban Dry Islands Effect in Beijing: Evidence from Satellite and Ground Observations. *Remote Sens.* **2022**, *14*, 809. [CrossRef]

5. Chen, L.; Frauenfeld, O.W. Impacts of urbanization on future climate in China. *Clim. Dyn.* **2016**, *47*, 345–357. [CrossRef]
6. Tsoka, S.; Tsikaloudaki, A.; Theodosiou, T. Analyzing the ENVI-met microclimate model's performance and assessing cool materials and urban vegetation applications—A review. *Sustain. Cities Soc.* **2018**, *43*, 55–76. [CrossRef]
7. Berardi, U.; Jafarpur, P. Assessing the impact of climate change on building heating and cooling energy demand in Canada. *Renew. Sustain. Energy Rev.* **2020**, *121*, 109681. [CrossRef]
8. Yang, J.; Jin, S.; Xiao, X.; Jin, C.; Xia, J.; Li, X.; Wang, S. Local climate zone ventilation and urban land surface temperatures: Towards a performance-based and wind-sensitive planning proposal in megacities. *Sustain. Cities Soc.* **2019**, *47*, 101487. [CrossRef]
9. Stewart, I.D.; Oke, T.R. Local Climate Zones for Urban Temperature Studies. *Bull. Am. Meteorol. Soc.* **2012**, *93*, 1879–1900. [CrossRef]
10. Yang, P.; Ren, G.Y.; Hou, W. Temporal-Spatial Patterns of Relative Humidity and the Urban Dryness Island Effect in Beijing City. *J. Appl. Meteorol. Climatol.* **2017**, *56*, 2221–2237. [CrossRef]
11. Unger, J.; Skarbit, N.; Gál, T. Absolute moisture content in mid-latitude urban canopy layer: Part 2: Results from Szeged, Hungary. *Acta Climatol.* **2018**, *51–52*, 47–56. [CrossRef]
12. Milelli, M.; Bassani, F.; Garbero, V.; Poggi, D.; von Hardenberg, J.; Ridolfi, L. Characterization of the Urban Heat and Dry Island effects in the Turin metropolitan area. *Urban CLim.* **2023**, *47*, 13. [CrossRef]
13. Deosthali, V. Impact of rapid urban growth on heat and moisture islands in Pune City, India. *Atmos. Environ.* **2000**, *34*, 2745–2754. [CrossRef]
14. Zhou, Y.; Wang, Z.H.; He, Y.Q.; Deng, S.H.; Leng, T.T. Features of Thermal and Humid Environments in Chenggong New District of Kunming. *J. Environ. Prot. Ecol.* **2020**, *21*, 390–404.
15. Yoshida, A.; Yasuda, R.; Kinoshita, S. Mobile Observation of Air Temperature and Humidity Distributions under Summer Sea Breezes in the Central Area of Osaka City. *Atmosphere* **2020**, *11*, 1234. [CrossRef]
16. Huang, X.J.; Song, J.Y.; Wang, C.H.; Chui, T.F.M.; Chan, P.W. The synergistic effect of urban heat and moisture islands in a compact high-rise city. *Build. Environ.* **2021**, *205*, 16. [CrossRef]
17. Yang, X.S.; Peng, L.L.H.; Chen, Y.; Yao, L.Y.; Wang, Q.Q. Air humidity characteristics of local climate zones: A three-year observational study in Nanjing. *Build. Environ.* **2020**, *171*, 16. [CrossRef]
18. Dunjic, J.; Milosevic, D.; Kojic, M.; Savic, S.; Luzanin, Z.; Secerov, I.; Arsenovic, D. Air Humidity Characteristics in "Local Climate Zones" of Novi Sad (Serbia) Based on Long-Term Data. *ISPRS Int. J. Geo-Inf.* **2021**, *10*, 810. [CrossRef]
19. Du, R.Q.; Song, J.Y.; Huang, X.J.; Wang, Q.; Zhang, C.; Brousse, O.; Chen, P.W. High-resolution regional modeling of urban moisture island: Mechanisms and implications on thermal comfort. *Build. Environ.* **2022**, *207*, 11. [CrossRef]
20. Shi, Y.; Zhang, Y. Urban morphological indicators of urban heat and moisture islands under various sky conditions in a humid subtropical region. *Build. Environ.* **2022**, *214*, 108906. [CrossRef]
21. Xiang, X.; Zhai, Z.; Fan, C.; Ding, Y.; Ye, L.; Li, J. Modelling future land use land cover changes and their impacts on urban heat island intensity in Guangzhou, China. *J. Environ. Manag.* **2024**, *366*, 121787. [CrossRef]
22. Luo, F.; Yang, Y.; Zong, L.; Bi, X. The interactions between urban heat island and heat waves amplify urban warming in Guangzhou, China: Roles of urban ventilation and local climate zones. *Front. Environ. Sci.* **2023**, *11*, 1084473. [CrossRef]
23. Shi, Y.R.; Xiang, Y.R.; Zhang, Y.F. Urban Design Factors Influencing Surface Urban Heat Island in the High-Density City of Guangzhou Based on the Local Climate Zone. *Sensors* **2019**, *19*, 20. [CrossRef] [PubMed]
24. Oke, T.R.; Mills, G.; Christen, A.; Voogt, J.A. *Urban Climates | | Atmospheric Moisture*; Cambridge University Press: Cambridge, UK, 2017; pp. 254–269. [CrossRef]
25. Cai, M.; Ren, C.; Xu, Y.; Dai, W.; Wang, X.M. Local Climate Zone Study for Sustainable Megacities Development by Using Improved WUDAPT Methodology—A Case Study in Guangzhou. *Procedia Environ. Sci.* **2016**, *36*, 82–89. [CrossRef]
26. Yang, X.S.; Yao, L.Y.; Jin, T.; Peng, L.L.H.; Jiang, Z.D.; Hu, Z.Y.; Ye, Y.H. Assessing the thermal behavior of different local climate zones in the Nanjing metropolis, China. *Build. Environ.* **2018**, *137*, 171–184. [CrossRef]
27. Chen, G.; He, M.J.; Li, N.; He, H.; Cai, Y.N.; Zheng, S.L. A Method for Selecting the Typical Days with Full Urban Heat Island Development in Hot and Humid Area, Case Study in Guangzhou, China. *Sustainability* **2021**, *13*, 320. [CrossRef]
28. Oke, T.R. Boundary Layer Climates. *Earth Sci. Rev.* **1987**, *27*, 265.
29. Bolton, D. The Computation of Equivalent Potential Temperature. *Mon. Weather. Rev.* **1980**, *108*, 1046–1053. [CrossRef]
30. Wexler, A. Vapor Pressure Formulation for Water in Range 0 to 100 °C. A Revision. *J. Res. Natl. Bur. Standards. Sect. A Phys. Chem.* **1976**, *80A*, 775–785. [CrossRef]
31. Kuttler, W.; Weber, S.; Schonnefeld, J.; Hesselschwerdt, A. Urban/rural atmospheric water vapour pressure differences and urban moisture excess in Krefeld, Germany. *Int. J. Climatol.* **2007**, *27*, 2005–2015. [CrossRef]
32. Mohajerani, A.; Bakaric, J.; Jeffrey-Bailey, T. The urban heat island effect, its causes, and mitigation, with reference to the thermal properties of asphalt concrete. *J. Environ. Manag.* **2017**, *197*, 522–538. [CrossRef]
33. Zhao, Z.Q.; He, B.J.; Li, L.G.; Wang, H.B.; Darko, A. Profile and concentric zonal analysis of relationships between land use/land cover and land surface temperature: Case study of Shenyang, China. *Energy Build.* **2017**, *155*, 282–295. [CrossRef]

34. Wang, Z.X.; Song, J.Y.; Chan, P.W.; Li, Y.G. The urban moisture island phenomenon and its mechanisms in a high-rise high-density city. *Int. J. Climatol.* **2021**, *41*, E150–E170. [CrossRef]
35. Stewart, I.; Oke, T. *Thermal Differentiation of Local Climate Zones Using Temperature Observations from Urban and Rural Field Sites*; University of British Columbia: Vancouver, BC, Canada, 2010.
36. Siu, L.W.; Hart, M.A. Quantifying urban heat island intensity in Hong Kong SAR, China. *Environ. Monit. Assess.* **2013**, *185*, 4383–4398. [CrossRef]
37. Fisher, J.B.; Malhi, Y.; Bonal, D.; Da Rocha, H.R.; De Araujo, A.C.; Gamo, M.; Goulden, M.L.; Hirano, T.; Huete, A.R.; Kondo, H.; et al. The land-atmosphere water flux in the tropics. *Glob. Change Biol.* **2009**, *15*, 2694–2714. [CrossRef]
38. Richards, K. Urban and rural dewfall, surface moisture, and associated canopy-level air temperature and humidity measurements for Vancouver, Canada. *Bound. Layer Meteorol.* **2005**, *114*, 143–163. [CrossRef]
39. Wang, Y.; Li, Y.; Chan, P.W. The Impact of Urbanization on Moisture Excess in Hong Kong. *Energy Procedia* **2015**, *78*, 3061–3065. [CrossRef]
40. Aljoumani, B.; Sanchez-Espigares, J.A.; Kluge, B.; Wessolek, G.; Kleinschmit, B. Analyzing Temporal Trends of Urban Evaporation Using Generalized Additive Models. *Land* **2022**, *11*, 508. [CrossRef]
41. Jim, C.Y.; Peng, L.L.H. Weather effect on thermal and energy performance of an extensive tropical green roof. *Urban For. Urban Green.* **2012**, *11*, 73–85. [CrossRef]
42. Hao, L.; Huang, X.L.; Qin, M.S.; Liu, Y.Q.; Li, W.H.; Sun, G. Ecohydrological Processes Explain Urban Dry Island Effects in a Wet Region, Southern China. *Water Resour. Res.* **2018**, *54*, 6757–6771. [CrossRef]
43. Luo, M.; Lau, N.-C. Urban Expansion and Drying Climate in an Urban Agglomeration of East China. *Geophys. Res. Lett.* **2019**, *46*, 6868–6877. [CrossRef]
44. Li, X.; Fan, W.; Wang, L.; Luo, M.; Yao, R.; Wang, S.; Wang, L. Effect of urban expansion on atmospheric humidity in Beijing-Tianjin-Hebei urban agglomeration. *Sci. Total Environ.* **2021**, *759*, 144305. [CrossRef]
45. Li, Y.; Fan, S.; Li, K.; Ke, K.; Dong, L. Contributions of urban green and blue spaces on local-scale climate in the core area of Beijing, China. *Urban Ecosyst.* **2023**, *26*, 1639–1650. [CrossRef]
46. Luo, Z.R.; Liu, J.H.; Zhang, Y.X.; Zhou, J.J.; Yu, Y.D.; Jia, R.T. Spatiotemporal characteristics of urban dry/wet islands in China following rapid urbanization. *J. Hydrol.* **2021**, *601*, 7. [CrossRef]
47. Luo, Z.R.; Liu, J.H.; Zhang, Y.X.; Zhou, J.J.; Shao, W.W.; Yu, Y.D.; Jia, R.T. Seasonal variation of dry and wet islands in Beijing considering urban artificial water dissipation. *NPJ Clim. Atmos. Sci.* **2021**, *4*, 9. [CrossRef]
48. Yang, X.; Xu, S.; Peng, L.L.H.; Chen, Y.; Yao, L. General air temperature and humidity features of local climate zones: A multi-city observational study in eastern China. *Urban Clim.* **2023**, *51*, 101652. [CrossRef]
49. Ribeiro, I.; Martilli, A.; Falls, M.; Zonato, A.; Villalba, G. Highly resolved WRF-BEP/BEM simulations over Barcelona urban area with LCZ. *Atmos. Res.* **2021**, *248*, 105220. [CrossRef]

buildings

MDPI

Article

The Patronization of Pluvial Flood Risk and Adaptation Among Tenant-Owned Housing Associations in Sweden

Mattias Hjerpe *, Erik Glaas and Sofie Storbjörk

Department of Thematic Studies—Environmental Change, and the Centre for Climate Science and Policy Research, Linköping University, SE 581 83 Linköping, Sweden; erik.glaas@liu.se (E.G.); sofie.storbjork@liu.se (S.S.)
* Correspondence: mattias.hjerpe@liu.se; Tel.: +46-708-840-470

Abstract: Pluvial floods are increasingly affecting urban areas worldwide. Despite growing media attention and clear owner responsibility for reducing climate-related risk for buildings in Swedish national adaptation policy, adaptation action remains slow. Understanding how different property owner categories view and act on flood risks is key for developing better incentive structures and support for accelerating adaptation action. While tenant-owned housing is a common form of housing tenure in Sweden, studies are lacking. This study enhances understanding of pluvial flood risk and adaptation views and actions by tenant-owned housing associations in two Swedish cities. It is based on assessments of 69 apartment buildings within eleven tenant-owned associations and semi-structured interviews with their eleven chairpersons. The study indicates that tenant-owned associations grossly underestimate their flood risks and responsibilities for climate adaptation, even though many buildings studied are at significant risk, and most associations have been impacted by floods, some severely and recurrently. The patronization of flood risk and responsibility for adaptation is attributed to several factors: underestimating risks and consequences, devaluing the benefit of one's own adaptation actions, lacking knowledge about climate adaptation measures for buildings, and (overly) generous insurance terms. The findings confirm low adaptation action among housing associations, even those with recurring floods, which is concerning given the strong reliance on property-owner adaptation in national adaptation policy.

Keywords: adaptation; climate risk; pluvial flood; tenant-owned housing; urban built environment

1. Introduction

With climate change, heavy precipitation is intensifying, increasing flood risks in built environments [1]. In many parts of the world, floods caused by heavy precipitation, here referred to as pluvial floods, have become more common and drastic, illustrated not least by the extensive damage and deaths caused by extreme rainfall in European cities in 2024. Such trends towards more powerful and devastating heavy rains are also found in Sweden [2,3], shown to cause significant damage to buildings and their residents [4,5]. This is demonstrated by the €160 million of insured damage costs as well as the impact on people and socially important operations in the Gävle pluvial flood in 2021 [6,7]. Consequently, adapting the built environment, including buildings and properties, to pluvial flood risks is becoming increasingly crucial.

Similar to other European countries [8], the main responsibility for climate adaptation of buildings and properties in Sweden lies with the property owner [9], making them key actors in the implementation of flood risk mitigation actions. However, despite this designated responsibility, studies indicate that few property owners have implemented climate adaptation measures [10–13]. Though this topic has generated increased academic attention, there are still gaps in understanding why relatively few climate adaptation measures have been implemented by property owners and how further action can be spurred [10,14–17]. The few existing studies on how Swedish property owners view and manage pluvial flood risk have mostly examined homeowners [10,13,18,19], and to a lesser extent property and construction companies [11,20]. Tenant-owned housing is a common form of housing tenure in Sweden, but studies of how this property owner views and responds to pluvial flood risk are lacking. Thus, how this property owner category views and acts on flood risks is still unclear, but it is important to study to build knowledge for developing more effective policies and incentive structures for climate adaptation of the built environment.

Related studies on flood risks and adaptation among homeowners generally conclude that homeowners are unfamiliar with evaluating pluvial flood risk, view it as rare and unfamiliar, and even in high-risk areas, tend to underestimate risk and rarely recognize it as something to consider [14–17]. A described reason for this underestimation is that it can be difficult to understand which factors generate flood risks in buildings, acting as a barrier to implementation [13]. While studies have identified various characteristics that contribute to the pluvial risk of a building, they have not found any statistically significant correlations [21], indicating high diversity in building stocks. However, several neighborhood and building characteristics have been demonstrated to contribute to a building's pluvial flood risk. Neighborhood characteristics include location in an anticipated floodplain or low-lying area [22], the capacity of the stormwater system/type of sewerage system [5,19], the share of impervious area, and topography [23]. In terms of vulnerability characteristics, studies have found that buildings with basements are flooded more frequently than others [22,24–26]. Additionally, vulnerability characteristics include the activities and people residing in the buildings [27] and the location of vital technical equipment [11]. Thus, though the combinations of characteristics that generate risk tend to be location-specific, it may still seem surprising that property owners usually do not connect obvious characteristics such as location in a low point to pluvial flood risk, but this seems to be the case with both private house owners [10,13,18,19] and property companies [6,7,11].

Studies have also analyzed the relationship between property owners' flood experience and willingness to adapt their property, generally finding that such experience does increase willingness [15]. However, it is also clear that flood experience does not necessarily result in adaptation, as shown in both international [15,16] and Swedish studies [10,13]. Studies of pluvial risk conclude that perceptions of one's capability to adapt to the risk influence perceptions of pluvial flood risk and adaptation behavior to a higher degree than perceptions of floods as a hazard [15,27]. This is often referred to as coping appraisal, which includes views on the effectiveness of available adaptation actions, one's ability to implement these actions, and their economic and emotional costs. Studies of Swedish homeowners' decision to adapt indicate that perceptions of adaptation actions are indeed important, highlighting the need for knowledge about how to adapt [10,13]. For instance, Grahn and Jaldell [10] found that homeowners who had adapted their property viewed themselves as knowledgeable about reducing flood risk. However, very few homeowners considered themselves knowledgeable about flood risks, and a majority stated that they were not sufficiently informed to evaluate whether the property-level flood risk reduction actions surveyed would reduce their flood risk, resulting in a low level of adaptation [10].

Hudson et al. [28] reviewed insurance schemes in European countries and found that existing schemes predominantly focus on recovery after weather events rather than proactively reducing climate risks and enhancing property owners' adaptive capacity. Studies have shown that it is challenging to find insurance schemes that successfully promote risk reduction [28,29]. These studies argue that such insurance schemes alone are insufficient, highlighting the need for research into the broader enabling environments and partnership arrangements required to promote proactive climate risk reduction [28,30]. Hudson et al. [31] characterized the Swedish insurance scheme for extreme weather as a semi-voluntary private market, where multiple extreme weather events are combined in a single policy, and the purchase of extreme weather event insurance is connected to a more commonly required product [28]. This implies that the home insurance of most Swedish homeowners covers floods, but the extent and type of damages covered vary, causing concern among Swedish homeowners who have experienced pluvial flooding [13].

A tenant-owned housing association—the property owner category addressed in this study—is a form of housing tenure where tenants co-own apartment buildings or small houses, regulated in Sweden by the Tenant-ownership Act [32]. By buying an apartment or semi-detached house, tenants get the right to reside in one of the apartments/houses and become members of the association [32]. Tenants are responsible for maintaining the interior of their apartments/houses, while the association is responsible for the property including common areas, water and electricity piping, roofs, facades, garages, and yards. Members pay a monthly fee to cover the operation and maintenance of the property. Tenant-owned housing associations are an increasingly common form of housing tenure in Sweden. In 2021, 42% of all apartments in apartment buildings in Sweden were in tenant-owned housing associations, a share that has increased over the last few decades [33]. Apart from making up a significant share of urban buildings in Sweden, tenant-owned housing associations are an interesting form of housing tenure to study because their tenants are laypeople, such as homeowners, but their buildings are similar to those owned by residential property companies. The board of the association is led by one of the tenants acting as chair, coordinating and interacting both with the property manager and the other tenants. This puts the chairperson in an important strategic position, mandated to lead the maintenance of the property in a changing climate. Researching the views and actions of such key actors in climate adaptation has been identified as pertinent in the climate adaptation literature, particularly regarding the role of private actors [20,34].

This study aims to enhance the understanding of pluvial flood risk in buildings and how flood risks and adaptation are viewed and acted upon within tenant-owned housing associations. The study is based on assessments of 69 apartment buildings within eleven tenant-owned associations and semi-structured interviews with the eleven chairpersons of these associations in two Swedish cities. The study is guided by three research questions:

- RQ1: To what extent and how are the assessed tenant-owned apartment buildings exposed and vulnerable to pluvial floods, and how are pluvial floods experienced and flood risks perceived by the interviewed chairpersons?
- RQ2: How have the analyzed tenant-owned apartment buildings been impacted by pluvial floods, and what are the reasons why these buildings have been flooded?
- RQ3: What actions did the associations take during and after a pluvial flood, and how are climate adaptation actions perceived by the interviewed chairpersons?

This article is structured as follows: Section 2 contains the materials and methods used. Section 3 presents the results divided into five sections, covering all three RQs. Section 4 discusses the results, and Section 5 concludes.

2. Materials and Methods

This study employed a desk study to assess pluvial flood risk of 69 apartment buildings within eleven tenant-owned associations and eleven semi-structured interviews with the chairperson of the respective association to analyze how flood risks are viewed and acted upon by the associations through the eyes of its chairperson. The study understands risk as the potential for adverse consequences caused by climate-related events, recognizing the dynamic interaction between hazards, exposure, and vulnerability [1,35]. The IPCC defines hazard as the potential occurrence of a physical event that may cause loss to property and infrastructure [36]. In this study, we researched how the housing associations' chairpersons perceived the potential occurrence of pluvial flood events (Section 3.1). In terms of exposure, studies of pluvial flood risks commonly suggest that exposure is influenced by the location of the building, local topography, and the portion of impervious surfaces. In this study, we report data on the pluvial flood exposure of the housing associations according to flood risk maps (Section 3.1), the chairpersons' views on what determines whether their property is considered at risk, and reasons why the buildings were flooded (Section 3.2) and mechanisms through which the buildings were flooded. Vulnerability encompasses both the sensitivity of the buildings, namely, to what extent the property owner views that the building will be impacted by a pluvial flood, and their adaptive capacity, namely, their ability to overcome the consequences [35]. The level of sensitivity of the buildings depends on the capacity of the sewage/drainage system, whether the building has a basement, the existence of openings, and the placement of indoor vital equipment [11]. In this study, a table contains occurrence and frequency of sensitivity characteristics such as ground floor apartments, basement stairs, and garages. Moreover, physical and financial consequences of the pluvial flood in focus are reported in Section 3.3. Sections 3.4 and 3.5 present actions performed during and after the pluvial flood as well as the housing association chairperson's views on adaptation actions, addressing issues of adaptive capacity.

Building on the author's ongoing collaboration with the umbrella organization for tenant-owned housing associations in southeastern and central Sweden, Riksbyggen, the study zoomed in on associations in the two mid-sized cities Linköping and Gävle. The cities have been affected by pluvial flooding to varying degrees, which were seen as an opportunity to capture a wide range of experiences and perspectives. In 2021, Gävle experienced national record-breaking precipitation [37] with return periods of about 3900 years, resulting in 6830 insurance claims exceeding €160 million in insured damage costs and impacts on welfare buildings [6,7]. Linköping has been affected by minor cloudburst events. In 2018, heavy precipitation flooded basements, a shopping area, and the central railway station.

Also, within the two cities, efforts were made to capture a heterogeneous set of associations. Thus, associations with varying degrees of pluvial flood exposure and sensitivity characteristics were identified through assessments of flood risk mappings of a 100-year rainfall event for Gävle [38] and Linköping [39] and through dialogues with staff from Riksbyggen. The pluvial flood exposure of the housing associations buildings was categorized using these flood risk maps as follows: high exposure >0.3 m water expected at least on one side of the building; medium exposure = 0.2–0.3 m water expected at least on one side of the building; and low <0.2 m water expected at least on one side of the building.

Through the dialogues, the association's size, building type, and building year were weighed in to ensure a heterogeneous sample. In both cities, Riksbyggen staff acted as door-openers [40], initially contacting the chairpersons to ask if they were willing to participate in the study. Most associations asked agreed to participate. The researcher then contacted the chairpersons to inform them about the study and to schedule and conduct the interview. The size of the sample allows for theme saturation in the interviews, following the recommendation in Wutich et al. [41].

Based on data retrieved from the websites of the eleven housing associations, the study covers 1954 apartments in 69 buildings owned by the eleven tenant-owned housing associations, ranging from 42 to 627 apartments per association, with construction years from 1949 to 2017. The sample also includes a variety of building types, such as high-rise buildings, three-story buildings, and terraced houses, which are common in both tenant-owned housing associations and the current built environment in Swedish urban areas.

In total, eleven interviews were conducted: six in Gävle (G1–6) and five in Linköping (L1–5). All interviewees were chairpersons of the board (Table 1), and, unintentionally, all were older men. The interview guide was based on questions about experiences with floods, their impacts, responses, and views on flood hazards, exposure, vulnerability, and responsibility. All interviewees provided informed consent before the study began, and they were free to withdraw at any time, and confidentiality was strictly maintained throughout the study.

Table 1. Background data and information about the tenant-owned housing associations and the interviews.

Interviewee/Housing Association	Details	Date	Length
Gävle 1 (G1)	47 apartments, 3 buildings, 4 stories, 2016	20220324	61 min
Gävle 2 (G2)	47 apartments, 8 buildings, terrace houses, 2 stories, 1973	20220325	140 min
Gävle 3 (G3)	42 apartments, 1 building, 6 stories, garage, basement, 1983	20220324	58 min
Gävle 4 (G4)	59 apartments, 3 buildings, 3 stories, basement, 1959	20220322	68 min
Gävle 5 (G5)	121 apartments, 3 buildings, 3–7 stories, basement, 1958	20220325	72 min
Gävle 6 (G6)	79 apartments, 3 buildings, high-rise, basement, 1966	20220324	54 min
Linköping 1 (L1)	129 apartments, 4 buildings, 4–6 stories, basement, 1985	20210414	62 min
Linköping 2 (L2)	380 apartments, 7 buildings, 3 stories with basement, 1953	20210416	62 min
Linköping 3 (L3)	627 apartments, 22 buildings, high-rise, garage, basements, 1955	20210423	105 min
Linköping 4 (L4)	284 apartments, 7 buildings, 3–5 stories, garages, 1949	20210424	117 min
Linköping 5 (L5)	139 apartments, 8 buildings, 2 stories, basements, 1963	20210427	46 min

All interviews were audio-recorded and transcribed. The transcripts were analyzed and coded based on the different elements of risk described above. An individual analysis of each transcript was followed by a comparative analysis to identify patterns, similarities, and differences. The main findings for each research question are presented in the results section below. The use of quotes from interviewees substantiates the analytical conclusions, and while striving to capture the main patterns in the material, individual perspectives are also presented to highlight distinctions.

3. Results

The results section presents findings from the desk study and interviews divided into five sections. The first section addresses RQ1, focusing on found pluvial flood risks for the assessed buildings and flood experiences and perceptions about flood risks among the interviewed chairpersons. The second and third sections address RQ2, reporting on reasons why the analyzed tenant-owned apartment buildings have suffered flood damage or not and on consequences of the experienced pluvial floods. Sections four and five address RQ3, presenting the actions taken by the tenant-owned apartment associations to reduce risk and damage during and after a pluvial flood and the chairpersons' perceptions about reactive and proactive climate adaptation actions.

3.1. Pluvial Flood Risks, Flood Experiences, and Perceptions About Flood Risks

Altogether, weighing both exposure and sensitivity components, the pluvial flood risk was quite high on average in the assessed buildings but differed both between and within the associations.

The exposure, according to flood risk maps, ranged from low in three tenant-owned housing associations, medium in two, and high in six (Table 2). In all associations but two, the exposure of individual buildings varied considerably within the association. In these associations, all buildings were either categorized as high (L3) or low (L1) exposure.

Table 2. Tenant-owned housing associations' pluvial flood risk experience, exposure, sensitivity characteristics, and expectation of pluvial flood in the future based on the desk study and interviews.

Housing Association	Exposure	BAS	BST	GF	Experience	Future Flood Expectation
G1	High			A, G	Single	No, very unlikely
G2	High	A		A, G	Multiple	Yes
G3	High	G			Single	Unlikely
G4	Medium	A, B	X		Single	Yes
G5	Low	B			Single	Yes
G6	Low	C		G	Multiple	Yes
L1	Low	G		A	Single	No, very unlikely
L2	High	G	X	A, P, G	Multiple	Yes
L3	High	B, C		A, G	None	No, very unlikely
L4	Medium	G	X		Single	No, very unlikely
L5	High	X	X	A, G	None	No, very unlikely
Tot		In 42 of 69 buildings	In 12 of 69 buildings	In 36 of 69 buildings		

GF = ground floor. GF A = ground floor apartment. GF G = Garage on the ground floor. GF P = Preschool on the ground floor. BAS = Basement. BAS A = Apartment in the basement. BAS B = Business space in the basement. BAS C = Common room in the basement. BAS G = Garage in the basement. BST = External basement stairs. High exposure >0.3 m of water expected at least on one side of the building. Medium exposure = 0.2–0.3 m of water expected at least on one side of the building. Low <0.2 m water expected at least on one side of the building.

Regarding the sensitivity of the tenant-owned housing associations' buildings, there were basements in 42 of 69 buildings across ten associations, with five of these basements housing apartments, businesses, or meeting rooms. Technical installations were always located in basements in buildings with basements. In total, there were apartments on the ground floor in 36 of 69 buildings across six associations, and one association also had a preschool on the ground floor. Additionally, 46 of 69 buildings in nine associations had garages, with four of these garages located in basements. Twelve buildings in four associations had external stairs leading to the basements, where only a few centimeters of surface water would be needed to flow down to the basement door, many of which were not waterproof.

In terms of personal experience with pluvial floods, the interviews established that nine of the eleven chairpersons, including all six from Gävle, had personal experience with pluvial flooding of buildings within their association. Three interviewees had multiple experiences, which they came to realize only after the most recent flood event. Interviewees with personal experiences described the pluvial flood event as dramatic and stressed that the event was unexpected.

All three interviewees with multiple pluvial flood experiences expected that their association's buildings would be at risk for pluvial floods in the future. Surprisingly, their answers still allowed for some uncertainty: 'Since it [pluvial flooding] occurred again, you really should think: Could it happen again?' (G2). This suggests that despite recurring pluvial floods, the chairperson still hesitated and said that future floods should not be taken for granted.

The views of future flood occurrence among interviewees with single experience diverged. Some interviewees considered their buildings at risk: 'So we're pretty sure that it [pluvial flood] will happen again' (G4), while others were more hesitant: 'If we look at the climate, it's not impossible' (G3), or believed pluvial floods were unlikely because they understand pluvial floods as very rare events: 'We have considered if there will be another flooding ... but hardly in our lifetime' (G1). Contrary to what would be expected after experiencing a pluvial flood, these interviewees believed it would not happen again, indicating risk denial.

Interviewees without personal pluvial flood experience did not consider their association's buildings at risk of future pluvial floods, even if they readily identified external basement stairs as the likely flood mechanism. This appeared to stem from a low perceived flood exposure, as illustrated by the following statement: 'I can't see that we currently are [at risk], unless the river Stångån rises ten meters' (L3). Interviewees without personal pluvial flood experience were informed about the location of the tenant-owned housing association's buildings on a pluvial flood risk map and the building's sensitivity characteristics, identified through the desk study, contributing to the pluvial flood risk of their buildings. This information did not influence the interviewees' view of the pluvial flood risk facing their association's buildings. Contrary to what the maps and desk study suggested, the interviewees referred to factors such as location in a heightened part of the city or on a slope, or very local topographical and building characteristics that were not accounted for on the map to support the view that their property would not be flooded in the future. Consequently, they did not consider their buildings at risk, regardless of the expected inundation levels represented on the pluvial flood maps.

Moreover, interviewees who connected pluvial flooding to climate change stated that this would make pluvial flood events more frequent in the future. However, since they considered pluvial flood events to be very rare, they struggled to make sense of and expressed doubts about what more frequent rainfall events would entail for something very rare. They reflected on the probability of a similar cloudburst to the one in Gävle in 2021: '...some prognosis said that there will be twice as many cloudbursts in the future... The problem with weather, though, is that you'll never know' (G2).

3.2. Reasons Why Buildings Have Been Flooded

Altogether, basements were flooded in all nine flood-damaged associations, and ground floors in four of them. While the interviewees could clearly describe which parts of the buildings were inundated, they struggled to ascertain the mechanisms through which the buildings were flooded and the exact locations where the water entered. The reported flood mechanisms spread evenly across surface openings, sewer backups, and other openings (Table 3). Six of the nine interviewees with personal pluvial flood experience contended that water entered the building through multiple mechanisms. Six interviewees reported surface openings, mainly entrance doors, whereas five identified that the flood occurred through backflow from sewage or stormwater pipes, mainly from drainage wells. Six interviewees noted other mechanisms, including from the building's foundation due to high groundwater or latent openings in roofs, on balconies, or windows. Interviews without personal pluvial flood experience identified external basement stairs as the likely flood mechanism in their buildings. Interviewees considered that improved knowledge about flood mechanisms would facilitate identification of actions to reduce the pluvial flood risk of their buildings. However, no interviewee reported any plans to analyze where the water entered their association's building(s).

Table 3. Pluvial flood mechanisms, consequences, and damage cost estimates.

Housing Association	Flood Mechanism	Consequences	Cost *
G1	Surface via entrances, elevator shafts, possibly foundation	Evacuation, ground floor apartments, electricity, storage	Total: €100,000 Own: 12,000
G2	Multiple: Surface via entrances and basement floor drains	Ground floor apartments and basement rooms, electricity, storage	Total: €500,000 Own: 20,000
G3	Multiple: Surface via entrances and garage door and top floor balconies	Ground floor apartments and basement rooms, elevators, storage, garage	Total and own: €20,000
G4	Multiple: Surface via entrances and drains outside basement garage	Basement apartment, business space, common room, garage	Total: €50,000 Own: 40,000
G5	Ground water through foundation	Basement common room incl. kitchen, sinkhole	Own cost: €10,000
G6	Multiple: Ground water through foundation and stormwater pipe burst	Basement common room incl. kitchen, storage, pipes	Do not know
L1	Multiple: roof windows and stormwater pipe burst/leakage	Ground floor apartments, garage, storage	-
L2	Multiple: Surface via entrances, windows, basement floor drains, and basement stair drains	Ground floor apartments (minor) and basement common room, garage, storage	-
L3	External basement stair drains	Do not think anything would be damaged	N/A
L4	Multiple: basement stair drains, garage door, foundation, and chimney	Garage, storage, possibly new basement apartment	-
L5	Some external basement stair drains	Do not think anything would be damaged	N/A

* The damage cost estimates reported here are either based on the costs that had occurred at the time of this study or the interviewee's best estimate.

Apart from these reasons, the interviewees with flood experience frequently referred to the insufficient capacity of the urban piped sewage or stormwater systems as a reason why their buildings were flooded. For example: 'Their [the municipality's] stormwater pipes cannot swallow the large quantity of water that came then [during the pluvial flood event], and we can't do anything about that. The municipality must deal with that'. (G4). This shows that tenant-owned housing association chairpersons tended to displace responsibilities for both the problem and undertaking adaptation responses to someone else, disavowing their own responsibility and role, as specified in national adaptation policy. It also indicates hopelessness and a sense of limited control over adapting their properties, i.e., they viewed that property-level adaptation would not make any difference to the building's pluvial flood risk.

3.3. *Impacts and Damages of the Pluvial Flood*

Contrary to the researchers' expectations, the interviewees with personal experience of pluvial flooding considered the impacts of the pluvial flood as tolerable, regardless of the magnitude and type of damage they had experienced. Even in the case where tenants had to be evacuated from their home for a long period of time, this was seen as tolerable, although interviewees contended that tenants did not want to evacuate and reported that some families decided to stay in their flooded apartments. Six interviewees reported damaged apartments on the ground floor or basement, viewed as aggravating the severity of the impact but still referred to as tolerable since there were no impacts on human health. No interviewee reported any personal injuries, but one interviewee was worried that tenants would be exposed to electricity in the basement, despite having advised them not to go there. Interestingly, one of the associations had recently constructed a basement apartment and, during the interview, realized that it would probably be damaged during a pluvial flood event. Since the interviewee had previously stated that impacts on people's homes were important to avoid during a pluvial flood, this caused some concern. This example also illustrates how low pluvial flood risk awareness can lead to higher risk. In six associations, the study found impacts on rented business spaces, referred to as unfortunate but unavoidable, and/or common rooms, referred to as tolerable. If a flooded building had an elevator, the elevator was commonly damaged, but this was referred to as tolerable damage because they were only out of operation for a day or two.

All interviewees with pluvial flood experience reported damage in basements, such as storage areas, garages, bike storage, laundry rooms, and common rooms, which were all referred to as minor damage. The flood mainly caused material damage to floor and wall construction materials, along with furniture and tenants' possessions in basement storage. Destroyed possessions were referred to as minor damage and were mostly viewed as a question of how to dispose of the rubbish. This is understandable since these damages are covered by the tenants' home insurance. However, two interviewees reported that damages to possessions had caused more severe impacts on already vulnerable tenants, including a woman in grief who had not cleaned out the properties of the deceased.

The study also found long-term consequences. For example, a hair salon located in a basement was damaged and restored, but the hairdresser noticed a bad smell afterward, triggering a new round of moisture measurement and investigation. The interviewees also noted that tenants were eager to notify them about any bad smells or indications of mold, signaling a worry that the water damage might have long-term consequences. Consequently, the interviewees indicated that moisture measurement became a preemptive strategy for both compliance with insurance company requirements and taking precautions against future water-related damage.

Financially, the study found large differences between estimations of both total damage costs and insurance-covered costs (Table 3). Interviewees' estimates of total costs ranged from €10,000 to €500,000, and their estimates of costs to be covered by the tenant-owned housing association ranged from €10,000 to €40,000. These cost estimates, however, did not include costs covered by tenants and their home insurance, which are likely to be considerable, particularly in cases where apartments were damaged. All interviewees with personal experience of floods referred to costs as manageable and not causing any financial risk for the association. Interviewees emphasized that, due to a requirement to set aside maintenance funds and a high annual turnover, the financial solvency of their association was high. As one chairperson stated, 'Even if we suffer a million in damage, it's not much compared with our annual budget of 8 to 9 million' (L4). All tenant-owned housing associations housing cooperatives were insured, adding to their economic solvency and significantly reducing the cost covered by the tenant-owned housing associations that were significantly damaged.

The two interviewees without pluvial flood experience did not think anything important would be damaged.

3.4. Actions During and After the Pluvial Flood

Interviewees with flood experience clearly described that the handling of the pluvial flooding was very time-consuming, including initiating and maintaining contact with insurers, construction companies, service providers, and worried tenants. This created a stressful situation for the chairpersons, also seen as a minor issue. The interviewees noted that the fact that most people were working from home due to the COVID-19 pandemic facilitated the handling of the pluvial flood: 'It was very lucky that most board members were working from home. Corona was here a blessing in disguise' (G2).

Interviewees with pluvial flood experience all responded by removing the water, dehumidifying, and removing damaged building material, seen as comprehensive, protracted, and time-consuming, albeit uncomplicated (Table 4). Due to the many damaged buildings in the Gävle pluvial flood event, the availability of dehumidifiers and pumps was severely restricted, protracting the response time.

Table 4. Pluvial flood actions performed by the housing associations during and after the flood event divided into reactive and proactive responses.

Housing Association	Actions During the Flood	Reactive Actions	Proactive Actions
G1	Pumping, dehumidify, remove building material.	Temporary barriers considered but dismissed	Sensitivity-reduction
G2	Pumping, dehumidify, remove building material.	Backflow valves considered but dismissed	Sensitivity-reduction, possibly assessing risks systematically
G3	Dehumidify, remove building material.	Balcony drainage, watertight entrance doors	No, do all we can and is outside our own control
G4	Pumping, dehumidify, remove building material.	-	Sensitivity-reduction, possibly watertight entrance doors
G5	Pumping, dehumidify, remove building material and furniture.	-	No, done all we can, sensitivity-reduction
G6	Pumping, dehumidify, remove building material and furniture.	-	Possibly assessing groundwater flood risk
L1	Dehumidify, remove building material.	Stormwater pipes and drains, entrance roofs, sealed roof windows	Never thought about it and no need
L2	Dehumidify, remove building material.	Improved infiltration, basement stair gutters, waterproof windows, greenery. Considered backflow valves but dismissed. Considered waterproof entrance, basement and garage doors but not allowed.	Discouraged by cultural heritage restrictions and outside one's own control
L3	-	Restored surface slope.	No need to reduce pluvial flood risk
L4	Dehumidify	Improve drainage; plugged garage drains. Chimneys planned whilst roofs are to be renovated. Considered basement stair roofs but dismissed	No need to reduce pluvial flood risk
L5	-	-	No need to reduce pluvial flood risk

Regarding reactive actions to rebuild damaged parts of the buildings, four interviewees stated that they had corrected one or more particular mechanisms observed during the pluvial flood, including sealing windows, waterproofing doors and windows, and plugging basement drains.

The study also found that four associations had considered but dismissed actions frequently promoted as viable by insurance companies, such as backflow valves, temporary flood barriers, and entrance door roofs. Disregard was motivated by a perceived low effectiveness or being unaesthetic. One association had considered a range of reactive actions but was unable to implement them due to cultural heritage rules.

The study found that four associations had taken action to reduce the building's sensitivity, i.e., decrease damage in case of a future flood event. The sensitivity-decreasing actions, such as using inorganic materials and paints, did not happen automatically and were only implemented because they were suggested by an employee of the cooperative's property manager, which most tenant-owned housing associations have contracted and regarded as a trusted source of knowledge.

Further, no interviewee reported that they had performed any actions to systematically assess and proactively reduce the pluvial flood risk of their buildings. In terms of planned adaptation actions, at the time of the study, i.e., at least ten months after the pluvial flood, only three interviewees, two of whom had multiple experiences, stated that they were considering assessing the pluvial flood risk of their buildings in the future. For one of them, the assessment would only consider a specific type of flood risk. Two interviewees who were expecting future flooding of their property stated that they believe they have done all property-level adaptation action they could do, referring to further adaptation action being constrained by other actors, mainly the municipality. Three interviewees with flood experience reported no plans, stating that they had never given pluvial flood risk reduction a thought because they did not expect any future flooding of their property. Evidently, the two interviewees without pluvial flood experience, despite high pluvial flood exposure according to flood risk maps, saw no need for actions to reduce pluvial flood risk.

3.5. Views of Reactive and Proactive Adaptation Actions

The actions implemented by the tenant-owned housing associations are predominantly reactive. According to the interviews, this reactive approach aligns with how their association responds to other building and technical equipment failures. However, such a reactive approach, without a learning ambition, clearly has its limitations. It neither facilitates correcting other similar failures nor understanding their root causes. The interviewees referred to reactive actions as uncomplicated compared to preventive actions and generally expressed a strong sense of control over reactive actions, believing them to be sufficient to protect their buildings from pluvial flood damage in the future.

In contrast to the interviewees' favorable view of reactive actions, their statements clearly indicate that preventive actions in general, and actions to reduce pluvial flood risk in particular, were highly unlikely to occur. Interviewees agreed that preventive actions lacked a clear purpose or benefit, illustrated by the following statement: 'It's not like we'll dig up some pipes in preventive purpose to prepare for the future. We've never done that and have no plans to do that either' (L4). The way the interviewees discussed preventive actions indicates both a lack of awareness of how pluvial flood risks could be reduced and a fear that the effectiveness of such actions would be severely constrained, as they depend on complementary actions outside the property boundary being implemented by other actors, mainly municipal departments. As one chairperson stated, 'But I also think that we have an external influence, where we can't affect. Ultimately it [the responsibility] must be placed on the municipality, in my view... we can't change that. We depend on the municipality' (L2). The above is particularly important in view of the expectations in the Swedish national climate adaptation policy, which clearly assigns the responsibility for climate adaptation in the urban built environment to property owners [7,12].

Another possible explanation of the interviewees' unfavorable view of proactive actions could be that they tended to view preventive actions very narrowly, predominantly involving the enlarged capacity of sewage and stormwater pipes. This points to limited knowledge about what adaptation to pluvial flood risk could entail and, consequently, what actions were being considered. This is very different from the preventive, often nature-based, adaptation actions discussed in urban adaptation literature. The interviewees viewed pipe-related actions as invisible for the tenants, very costly, and substantially more expensive compared to their perceived minuscule benefits. As one chairperson stated,

'Underground actions will not be seen. Five years from now everybody will have forgotten about them. Investing many millions that nobody saw, the fee was increased, but nobody felt that they benefited' (L4). The visibility of actions was clearly seen as important for the interviewees, exemplified with: 'If you mount photovoltaic panels, they are visible externally, everybody sees them... it feels quite good in everybody's guts when you look at them' (L4).

The Interviewees also described a lower degree of control over preventive actions as they fall on the edge of, or outside, the tenant-owned housing association board's mandate compared with reactive actions. These would need to be decided at the tenant-owned housing associations' annual meeting, involving all tenants. The chairpersons emphasized that convincing tenants is very difficult.

4. Discussion

Based on assessments of pluvial risk characteristics in 69 buildings and interviews with chairpersons in eleven housing associations in two Swedish cities, the study provides five insights into how this previously understudied category of property owner views and acts upon flood risks and adaptation of buildings.

Firstly, the study showed that none of the eleven tenant-owned housing associations had adapted their buildings to pluvial floods, regardless of the buildings' level of pluvial flood exposure, sensitivity characteristics, and experienced impacts. This finding aligns with the low uptake of property-level adaptation actions found among homeowners [10,13] and property companies [7,11] in Sweden [12] and abroad [15,16,24]. Many of the tenant-owned apartment buildings examined were at risk for pluvial floods and, consequently, in need of adaptation. This is supported by experiences of flooded buildings in nine of the eleven associations studied, three of them several times, as well as high exposure indicated by expected surface water depth during a hundred-year event exceeding 0.30 m at least in parts of the buildings in six of them. Moreover, the study also found building characteristics associated with pluvial flood risk in all tenant-owned housing associations, such as apartments or businesses in basements in five of the eleven associations. Further, buildings in six associations were flooded through multiple flood mechanisms, indicating a need for property-level adaptation actions. Comparing pluvial flood exposure from the desk study with tenant-owned housing association chairpersons' views on future flood expectation indicates that they heavily underestimate pluvial flood occurrence, particularly among those with single or no personal flood experience. This underestimation has also been found in studies of other property owner categories in Sweden and abroad [10,15,16]. The study also suggests that merely making flood risk maps available is unlikely to influence the housing association chairpersons' pluvial flood risk perception and that vigorous action is needed to get this type of property owner to implement climate adaptation measures on/in their buildings and yards.

Secondly, contrary to the researchers' expectations, the study established that regardless of the type and magnitude of experienced damage during pluvial flood events, chairpersons viewed the damage as tolerable. This view appears to serve as a further reason why the tenant-owned housing association should not engage more with adaptation, similar to a rebound effect [16,17]. This reaffirms the importance of the experienced severity of impacts, rather than experience itself, as an influential factor for shaping individuals' views of pluvial flood risk [16]. The study suggests that impacts on tenants' health would likely trigger more engagement with climate adaptation of their buildings. Although evacuation of tenants and flooded business spaces were viewed by the chairpersons as impacts to be avoided, experiences of these damages did not trigger any engagement with proactive climate adaptation in the three associations experiencing these impacts. The underestimation of pluvial flood risk, combined with viewing pluvial flood damage as tolerable, indicates a very low perception of pluvial flood threat.

Thirdly, the financial impact of the pluvial flood was also seen as tolerable, which appears due to the fact that all tenant-owned housing associations were insured, demonstrating a high penetration of flood insurance, similar to other property owner categories in Sweden, but unlike in some other countries [16,24]. The study also found that insurance covered a large part of the damage costs, and the tenants' home insurance covered most of the cost of damaged possessions, even if the insurer's final decision was not made in some cases at the time of the study, ten months after the pluvial flood event. Additionally, no interviewee mentioned that their associations' insurance company had raised concerns about or demanded proactive adaptation during restoration, thus focusing solely on restoring rather than reducing pluvial flood exposure and sensitivity after the flood event, which is in line with experiences of insurance company activity in studies of other types of property owners in Sweden [7,11]. This indicates that insurance conditions and insurance company actions had no or very limited governing effect on this property owner's attitudes or actions regarding flood risks (cf [28,29]). This is noteworthy considering that insurance companies in many other countries have significantly increased their requirements on insurance takers, increased premiums, and/or separated flood risk insurance [42], even if the general conclusion is that insurance schemes alone are insufficient [28,31].

Fourthly, the study found that tenant-owned housing association chairpersons with personal experience of floods viewed pluvial flood impacts as financially and emotionally less significant than homeowners have done [13,18]. Even though the study found substantial damage costs, the chairpersons deemed these as minor due to their association's high economic solvency, high share of damage costs covered by insurance, and relatively low costs per tenant. This contrasts with the more substantial economic damage [10,13] and worry about the financial consequences [13] reported by homeowners. Moreover, the chairpersons in this study referred to themselves and their tenants as having high economic solvency, similar to what Jarnkvist [18] found among wealthy white male homeowners positioning themselves as standing above climate risks, or what O'Brien et al. [43] referred to as complacency to climate risk. This could be one more reason for the housing associations' low engagement with climate adaptation of their buildings.

Fifthly, the study found support for both a low sense of control of adaptation actions and a low perceived need for climate adaptation. Consequently, very few associations had adapted their buildings, and interviewees provided many arguments for why adaptation is unnecessary or ineffective for them. This suggests a weak sense of mandate and perceived capacity to control or make a difference in urban adaptation. This appears underpinned not only by a lack of knowledge of adaptation actions, as also found in studies of Swedish homeowners [10], but also by the chairpersons' views that many property-level flood reduction actions would be unsuitable for their association due to poor aesthetics and low efficiency. Here, the study found a very restrictive understanding of proactive adaptation actions, primarily seen to involve expensive investments in the pipe-bound sewage or stormwater system. This must be viewed in light of the wider palette of proactive adaptation actions available, often involving nature-based solutions or sustainable stormwater measures, including green roofs, rainbeds and rain gardens, open streams, permeable pavements, and vegetated swales [44,45], which no interviewee referred to. Such measures can reduce pluvial flood impacts and associated economic losses [45] and are also likely to be attractive to tenant-owned housing associations due to their visibility and aesthetics, pointing to a knowledge gap similar to what Grahn and Jaldell [11] found among Swedish homeowners. Moreover, regarding available adaptation actions, the study found that chairpersons' understanding of adaptation actions was shaped by their favored reactive approach to fixing problems after they had occurred. This approach, when practiced without reflection, fails to result in learning and remedies the symptoms rather than the causes of the pluvial flood risk. The interviewed housing association chairpersons also believed that the cost of adaptation would overwhelmingly exceed its benefits and that the risk to their buildings is foremost influenced by factors outside their own control, similar to findings in studies of homeowners [10,13,18]. This highlights the need to make information about proactive adaptation actions and the role of property owners in urban climate adaptation more widespread.

5. Conclusions

Pluvial floods are increasingly affecting urban areas worldwide. Despite this, available evidence suggests that adaptation actions among property owners remain slow, in spite of clear principles of owner responsibility for reducing climate-related risks for buildings and central roles attributed to property owners in national and local adaptation strategies. A greater understanding of how property owners view and act on flood risks has thereby been requested. In Sweden and many other countries, though, there are yet few studies focusing on how other property owner categories than homeowners views and act on flood risks and adaptation.

This study hones in on tenant-owned housing, which is an increasingly common form of housing tenure in Sweden, making up 42% of all apartments in apartment buildings in Sweden in 2021. The study set out to examine to what extent and how the assessed tenant-owned apartment buildings were exposed and vulnerable to pluvial floods, as well as how the interviewed chairpersons have experienced and viewed pluvial flood risks and adaptation. The study found that a high share of the tenant-owned housing association buildings studied were exposed to pluvial floods, supported both by desk study data indicating exposure and sensitivity characteristics and that nine of the eleven associations had been flooded at least once. Despite this, few housing associations stated a need for proactively adapting their buildings, indicating that they heavily underestimate pluvial flood risks, particularly those with single or no personal flood experience.

The study also found that the tenant-owned apartment buildings were impacted by pluvial floods not only through considerable and prolonged actions to dehumidify and rebuild the damaged parts of the buildings but also through flooded apartments, business spaces, garages, and common rooms, as well as the evacuation of tenants and damage costs of up to €500,000. Regardless of the damage type and magnitude, the chairpersons viewed the damage as tolerable, serving as a reason why they do not have to engage more with adaptation. The underestimation of pluvial flood risk, combined with viewing pluvial flood damage as tolerable, demonstrates a very low perception of pluvial flood threat and, consequently, a lack of reason to engage with climate adaptation.

Moreover, the study found support for both a low sense of control of adaptation actions and a low perceived need for climate adaptation, suggesting a very low perception of coping appraisal manifested as a weak sense of mandate and capacity to control or make a difference in urban adaptation. This, too, provides another reason for their low engagement with climate adaptation.

The study suggests that adaptation of tenant-owned apartment buildings is a complex task for laypeople such as the chairpersons in this study. It was found that sensitivity-reducing property-level actions were only implemented when suggested by a trusted source of expertise, which, in this study, turned out to be a professional property manager. Two further actors were found to have the potential to influence housing associations' adaptation engagement: municipalities and the insurance industry. Municipalities are recognized as key actors since sewerage departments have a mandate to manage urban stormwater and are attributed responsibility for deficient capacity, aggravating the flood risk of buildings in certain locations. To spur property-owner adaptation, it is found vital that municipalities inform about their adaptation actions, since property owners otherwise remain unaware. The insurance industry had, in this study, no or very limited governing effect on this property owner's attitudes or actions regarding flood risks since their actions predominantly targeted recovery rather than proactive reduction of pluvial flood risk. This is noteworthy considering that insurance companies in many other countries have significantly increased their requirements on insurance takers, increased premiums, and/or separated flood risk insurance [38]. This points to a potential for insurers to increase the incentives for property owner adaptation.

To sum up, the property owner category in focus for this study—tenant-owned housing associations—largely seems to underestimate pluvial flood risks for their property, even the ones impacted repeatedly. As a rule, they do not implement any proactive adaptation measures. This is alarming given that tenant-owned housing is a common form of housing in Sweden and that property owners have the main responsibility for adapting buildings in a changing climate. Powerful actions from the state, municipalities, insurance companies, and umbrella organizations appear needed both to steer towards a different course of action and to educate the actors about factors contributing to pluvial flood risk, common types of damages, the significant actions needed to restore damage, effective climate adaptation measures, and co-benefits of adaptation. Even though the study concerns tenant-owned housing associations in Sweden, the findings would also be relevant for other property owner categories, especially homeowners who also are laypeople such as the housing association chairpersons studied here. Moreover, since the apartment buildings studied do not differ significantly from such buildings in other cities, especially in northern Europe and North America, the findings related to flood mechanisms, types of damage occurred, actions during the pluvial flood, and reactive actions are likely to be more widely applicable.

Based on the findings in this study, future research should focus on ways to spur housing associations' engagement with climate adaptation and evaluate the effectiveness of various property-level adaptation actions. Here, studies of nature-based solutions would seem particularly promising due to their visibility and aesthetics. Moreover, studies examining to what extent adaptation action can be triggered by trusted sources of expertise are also needed.

Author Contributions: Conceptualization, M.H., E.G. and S.S.; Methodology, M.H.; Formal Analysis, M.H. and E.G.; Investigation, M.H.; Project Administration, M.H., Writing—Original Draft Preparation, M.H., E.G. and S.S.; Writing—Review and Editing, M.H., E.G. and S.S.; Funding Acquisition, M.H., E.G. and S.S. All authors have read and agreed to the published version of the manuscript.

Funding: This research was funded by Länsförsäkringsbolagens Forskningsstiftelse grant number P1.19.

Data Availability Statement: The raw data supporting the conclusions of this article will be made available by the corresponding author on request.

Acknowledgments: The authors are grateful to Angelica Zetterlund, Johanna Österås, and Denise Wallman, and to staff at Riksbyggen Gävleborg and Östergötland.

Conflicts of Interest: The authors declare no conflicts of interest.

References

1. Intergovernmental Panel on Climate Change (IPCC). Summary for Policymakers. In *Climate Change 2022: Impacts, Adaptation and Vulnerability. Contribution of Working Group II to the Sixth Assessment Report of the Intergovernmental Panel on Climate Change*; Pörtner,

H.O., Roberts, D.C., Tignor, M., Poloczanska, E.S., Mintenbeck, K., Alegría, A., Craig, M., Langsdorf, S., Löschke, S., Möller, V., et al., Eds.; Cambridge University Press: Cambridge, UK; New York, NY, USA, 2022; pp. 3–33.

2. Swedish Meteorological and Hydrological Institute (SMHI). *Climate Extremes for Sweden: State of Knowledge and Implications for Adaptation and Mitigation*; SMHI: Norrkoping, Sweden, 2019. [CrossRef]
3. Knös, D.; Karagiorgos, K.; Haas, J.; Blumenthal, B.; Nyberg, L.; Halldin, S. Cloudburst-disaster modelling: A new open-source catastrophe model. *Int. J. Disaster Risk Reduct.* **2022**, *67*, 102679. [CrossRef]
4. Nyberg, L.; Hakkarainen, H.; Blumenthal, B.; Moberg, J.-O. *Konsekvenser av Sommarskyfall i Sverige under Åren 2009–2018 (Consequences of Summer Cloudbursts in Sweden in Years 2009–2018, in Swedish)*; Centrum för klimat och säkerhet: 2019:2; Karlstads Universitet: Karlstad, Sweden, 2019.
5. Sörensen, J.; Mobini, S. Pluvial, urban flood mechanisms and characteristics—Assessment based on insurance claims. *J. Hydrol.* **2017**, *555*, 51–67. [CrossRef]
6. Finansinspektionen. *Försäkringsföretagen och översvämningarna i Gävleborg (Insurance Companies and the Floods in Gävleborg, in Swedish)*; Report 12; Finansinspektionen: Stockholm, Sweden, 2023.
7. Glaas, E.; Hjerpe, M.; Storbjörk, S. The 2021 extreme rainfall in Gävle, Sweden: Impacts on municipal welfare services and actions towards more resilient premises and operations. *Hydrol. Res.* **2021**, *55*, 431–443. [CrossRef]
8. Thaler, T.; Hartmann, T. Justice and flood risk management: Reflecting on different approaches to distribute and allocate flood risk management in Europe. *Nat. Hazards* **2016**, *83*, 129–147. [CrossRef]
9. Swedish Government. *Nationell strategi för klimatanpassning (National Climate Adaptation Strategy, in Swedish)*; Government Bill 2017/18:163; Swedish Government: Stockholm, Sweden, 2018.
10. Grahn, T.; Jaldell, H. Households (un)willingness to perform private flood risk reduction—Results from a Swedish survey. *Saf. Sci.* **2019**, *116*, 127–136. [CrossRef]
11. Hjerpe, M.; Glaas, E.; Hedenqvist, R.; Storbjörk, S.; Navarra, C. A systematic approach for assessing climate vulnerabilities and adaptation options in large property portfolios: Influences on property owners' transformative capacity. *IOP Conf. Ser. Earth Environ. Sci.* **2020**, *588*, 032044. [CrossRef]
12. National Climate Adaptation Expert Council. *Första Rapporten från Nationella Expertrådet för Klimatanpassning (First Report from the Swedish Expert Council on Climate Adaptation, in Swedish)*; National Climate Adaptation Expert Council: Norrkoping, Sweden, 2022.
13. Malmqvist, A.; Hjerpe, M.; Glaas, E.; Lundgren, T.; Gyberg, P.; Storbjörk, S. I am affected—This is for the municipality to solve: An interview study with Swedish villa owners affected by floods from cloudbursts [Jag drabbas—Det här får kommunen lösa: En intervjustudie med svenska villaägare som påverkats av översvämningar från skyfall, in Swedish]. *Sociol. Forsk.* **2024**, *60*, 275–298. [CrossRef]
14. Bichard, E.; Kazmierczak, A. Are homeowners willing to adapt to and mitigate the effects of climate change? *Clim. Change* **2012**, *112*, 633–654. [CrossRef]
15. Bubeck, P.; Botzen, W.J.W.; Aerts, J.C.J.H. A review of risk perceptions and other factors that influence flood mitigation behavior. *Risk Anal.* **2012**, *32*, 1481–1495. [CrossRef]
16. Environment Agency. *Applying Behavioural Insights to Property Flood Resilience*; Research Report FRS17171 commissioned by the Joint Flood and Coastal Erosion Risk Management Research and Development Programme; Environment Agency: Bristol, UK, 2020.
17. Thistlethwaite, J.; Brown, C.; Henstra, D.; Scott, D. How flood experience and risk perception influences protective actions and behaviours among Canadian homeowners. *Environ. Manag.* **2018**, *61*, 197–208. [CrossRef]
18. Jarnkvist, K. Intersectional perspectives of house owner narratives on climate risks. *J. Risk Res.* **2019**, *23*, 1211–1224. [CrossRef]
19. Mobini, S.; Nilsson, E.; Persson, A.; Becker, P.; Larsson, R. Analysis of pluvial flood damage costs in residential buildings—A case study in Malmö. *Int. J. Disaster Risk Reduct.* **2021**, *62*, 102407. [CrossRef]
20. Storbjörk, S.; Hjerpe, M.; Isaksson, K. "We cannot be at the forefront, changing society". Exploring how Swedish property developers respond to climate change in urban planning. *J. Environ. Policy Plan.* **2018**, *20*, 81–95. [CrossRef]
21. Gradeci, K.; Labonnote, N.; Sivertsen, E.; Time, B. The use of insurance data in the analysis of Surface Water Flood events—A systematic review. *J. Hydrol.* **2019**, *568*, 194–206. [CrossRef]
22. Gandini, A.; Garmendia, L.; Prieto, I.; Alvarez, I.; San-José, J.T. A holistic and multi-stakeholder methodology for vulnerability assessment of cities to flooding and extreme precipitation events. *Sustain. Cities Soc.* **2020**, *63*, 102437. [CrossRef]
23. Torgersen, G.; Rød, J.K.; Kvaal, K.; Bjerkholt, J.T.; Lindholm, O.G. Evaluating flood exposure for properties in urban areas using a multivariate modelling technique. *Water* **2017**, *9*, 318. [CrossRef]
24. Rözer, V.; Müller, M.; Bubeck, P.; Kienzler, S.; Thieken, A.; Pech, I.; Schröter, K.; Buchholz, O.; Kreibich, H. Coping with pluvial floods by private households. *Water* **2016**, *8*, 304. [CrossRef]
25. Papathoma-Köhle, M.; Schlögl, M.; Fuchs, S. Vulnerability indicators for natural hazards: An innovative selection and weighting approach. *Sci. Rep.* **2019**, *9*, 15026. [CrossRef]

26. Godfrey, A.; Ciurean, R.L.; van Westen, C.J.; Kingma, N.C.; Glade, T. Assessing vulnerability of buildings to hydro-meteorological hazards using an expert based approach—An application in Nehoiu Valley, Romania. *Int. J. Disaster Risk Reduct.* **2015**, *13*, 229–241. [CrossRef]
27. Othmer, F.J.; Becker, D.; Schulte, L.M.; Greiving, S. A methodological approach to municipal pluvial flood risk assessment based on a small city case study. *Sustainability* **2020**, *12*, 10487. [CrossRef]
28. Hudson, P.; De Ruig, L.T.; de Ruiter, M.C.; Kuik, O.J.; Botzen, W.J.W.; Le Den, X.; Persson, M.; Benoist, A.; Nielsen, C.N. An assessment of best practices of extreme weather insurance and directions for a more resilient society. *Environ. Hazards* **2020**, *19*, 301–321. [CrossRef]
29. Surminski, S.; Oramas-Dorta, D. Flood insurance schemes and climate adaptation in developing countries. *Int. J. Disast. Risk Reduct.* **2014**, *7*, 154–164. [CrossRef]
30. Surminski, S.; Thieken, A. Promoting flood risk reduction: The role of insurance in Germany and England. *Earth's Future* **2017**, *5*, 979–1001. [CrossRef]
31. Hudson, P.; Botzen, W.J.W.; Aerts, J.C.J.H. Flood insurance arrangements in the European Union for future flood risk under climate and socioeconomic change. *Glob. Environ. Change* **2019**, *58*, 101966. [CrossRef]
32. SFS 1991:614. Bostadsrättslag (Tenant Ownership Act, in Swedish). Available online: https://www.riksdagen.se/sv/dokument-lagar/dokument/svensk-forfattningssamling/bostadsrattslag-1991614_sfs-1991-614 (accessed on 31 August 2023).
33. Statistics Sweden. Antal Lägenheter efter Hustyp, Upplåtelseform och År. Antal Lägenheter efter Hustyp, Upplåtelseform och År (Number of Apartments According to Building Type, Form of Tenure and Year, in Swedish). PxWeb (scb.se). Available online: https://www.statistikdatabasen.scb.se/pxweb/sv/ssd/START__BO__BO0104__BO0104D/BO0104T04/ (accessed on 29 August 2023).
34. Mees, H. Local governments in the driving seat? A comparative analysis of public and private responsibilities for adaptation to climate change in European and North-American cities. *J. Environ. Policy Plan.* **2017**, *19*, 374–390. [CrossRef]
35. Intergovernmental Panel on Climate Change (IPCC). *The Concept of Risk in the IPCC Sixth Assessment Report: A Summary of Cross-Working Group Discussions*; Reisinger, A., Howden, M., Vera, C., Garschagen, M., Hurlbert, M., Kreibiehl, S., Mach, K.J., Mintenbeck, K., O'Neill, B., Pathak, M., et al., Eds.; Intergovernmental Panel on Climate Change: Geneva, Switzerland, 2020.
36. Connelly, A.; Carter, J.; Handley, J.; Hincks, S. Enhancing the practical utility of risk assessments in climate change adaptation. *Sustainability* **2018**, *10*, 1399. [CrossRef]
37. Swedish Meteorological and Hydrological Institute (SMHI). 2021—Skyfall i Gävle (Cloudburst in Gävle, in Swedish). Available online: https://www.smhi.se/kunskapsbanken/hydrologi/historiska-oversvamningar/2021-skyfall-i-gavle-1.175548 (accessed on 20 August 2023).
38. Gästrike Vatten. *Teknisk Modelldokumentation: Ytvattenmodell Gävle (Technical Model Documentation: Surface Water Model Gävle, in Swedish)*; Uppdrag: 10329031; Gästrike Vatten: Gävle, Sweden, 2023.
39. Linköpings Kommun. *Skyfallskartering Linköpings Kommun (Cloudburst Map: Linköping Municipality, in Swedish)*; Uppdrag: 12803421; Linköpings Kommun: Linköping, Sweden, 2016.
40. Dalen, M. *Intervju som Metod (Interview as a Method, in Swedish)*; Gleerups: Falkenberg, Sweden, 2015.
41. Wutich, A.; Beresford, M.; Bernard, H.R. Sample sizes for 10 types of qualitative data analysis: An integrative review, empirical guidance, and next steps. *Int. J. Qual. Methods* **2024**, *23*, 16094069241296206. [CrossRef]
42. Botzen, W.J.W.; Van Den Bergh, J.C.J.M. Insurance against climate change and flooding in the Netherlands: Present, future, and comparison with other countries. *Risk Anal.* **2008**, *28*, 413–426. [CrossRef] [PubMed]
43. O'Brien, K.; Eriksen, S.; Sygna, L.; Naess, L.-O. Questioning complacency: Climate change impacts, vulnerability, and adaptation in Norway. *Ambio* **2006**, *35*, 50–56. Available online: https://www.jstor.org/stable/4315686 (accessed on 1 November 2023). [CrossRef] [PubMed]
44. Karamoutsou, L.; Veliou, E.; Bigum Kvernberg, E.; Psilovikos, A. Urban stormwater management: A sustainable approach. *Water* **2024**, *16*, 2089. [CrossRef]
45. Knapika, E.; Brandimarte, L.; Usher, M. Maintenance in sustainable stormwater management: Issues, barriers and challenges. *J. Environ. Plan. Manag.* **2024**, 1–27. [CrossRef]

buildings

MDPI

Article

Multi-Building Energy Forecasting Through Weather-Integrated Temporal Graph Neural Networks

Samuel Moveh [1], Emmanuel Alejandro Merchán-Cruz [1], Maher Abuhussain [2], Saleh Alhumaid [3], Khaled Almazam [4,*] and Yakubu Aminu Dodo [4,*]

1 Engineering Faculty, Transport and Telecommunication Institute, 1019 Riga, Latvia; moveh.s@tsi.lv (S.M.); merchan.e@tsi.lv (E.A.M.-C.)
2 Department of Civil and Environmental Engineering, College of Engineering and Computing in Al-Qunfudhah, Umm Al-Qura University, Mecca 24221, Saudi Arabia; maabuhussain@uqu.edu.sa
3 Department of Mechanical Engineering, University of Hail, Hail 81481, Saudi Arabia; s.alhumaid@uoh.edu.sa
4 Architectural Engineering Department, Najran University, Najran 66426, Saudi Arabia
* Correspondence: kaalmazam@nu.edu.sa (K.A.); yadodo@nu.edu.sa (Y.A.D.)

Abstract: While existing building energy prediction methods have advanced significantly, they face fundamental challenges in simultaneously modeling complex spatial–temporal relationships between buildings and integrating dynamic weather patterns, particularly in dense urban environments where building interactions significantly impact energy consumption patterns. This study presents an advanced deep learning system combining temporal graph neural networks with weather data parameters to enhance prediction accuracy across diverse building types through innovative spatial–temporal modeling. This approach integrates LSTM layers with graph convolutional networks, trained using energy consumption data from 150 commercial buildings over three years. The system incorporates spatial relationships through a weighted adjacency matrix considering building proximity and operational similarities, while weather parameters are integrated via a specialized neural network component. Performance evaluation examined normal operations, data gaps, and seasonal variations. The results demonstrated a 3.2% mean absolute percentage error (MAPE) for 15 min predictions and a 4.2% MAPE for 24 h forecasts. The system showed robust data recovery, maintaining 95.8% effectiveness even with 30% missing values. Seasonal analysis revealed consistent performance across weather conditions (MAPE: 3.1–3.4%). The approach achieved 33.3% better prediction accuracy compared to conventional methods, with 75% efficiency across four GPUs. These findings demonstrate the effectiveness of combining spatial relationships and weather parameters for building energy prediction, providing valuable insights for energy management systems and urban planning. The system's performance and scalability make it particularly suitable for practical applications in smart building management and urban sustainability.

Keywords: building energy prediction; temporal graph neural networks; spatial–temporal modeling; multi-building systems; weather integration

1. Introduction

The fast pace of urbanization and the growing complexity of buildings nowadays have led to a sharp increase in energy use around the world. The building sector is responsible for around 40% of global operational energy consumption, primarily due to heating, cooling, lighting, and other building services. Among these, lighting and appliances typically account for about 20–30%, varying by building type and region, while heating, ventilation, and air-conditioning (HVAC) systems often constitute the largest portion, contributing approximately 50–60% of a building's operational energy use [1–3]. Accurate energy prediction systems serve as essential enablers for sustainability initiatives by facilitating optimal resource allocation, enabling proactive energy management, and supporting smart building operations [4,5]. These systems are particularly crucial for integrating renewable energy sources, implementing effective demand response programs, and identifying opportunities for energy conservation, thereby directly contributing to broader sustainability goals.

Recent progress in deep learning has transformed building energy prediction approaches, offering new opportunities to understand complex patterns in energy data [6,7]. While traditional statistical analysis and regression models have demonstrated success in many building energy applications, capturing complex non-linear interactions between multiple factors (such as weather conditions, occupancy patterns, and building characteristics) can be challenging with these approaches alone. Recent studies suggest that for scenarios involving multiple buildings with strong spatial–temporal dependencies, advanced machine learning approaches may offer additional modeling capabilities, though simpler methods remain valuable for many use cases [8].

Recent advancements in building energy prediction have been demonstrated in studies that leverage novel deep learning and graph-based techniques to capture complex spatial–temporal interactions. For instance, Hu et al. [9] proposed a graph convolutional network-based approach that integrates inter-building dependencies for urban energy forecasting. Similarly, Simeunović et al. [10] presented a spatio-temporal graph neural network model for indoor environment prediction and optimal control in central air-conditioning systems, demonstrating improved accuracy over traditional methods. Moreover, recent work by Liao et al. [11] on hybrid graph convolutional networks (GCNs)–Long Short-Term Memory (LSTM) models for short-term renewable power prediction further illustrates the effectiveness of integrating graph-based spatial learning with temporal models. These studies underscore the necessity and timeliness of integrated frameworks that can simultaneously model spatial and temporal dependencies in building energy prediction.

The emergence of temporal graph neural networks (TGNNs) offers novel approaches to model these complex interactions while considering both spatial and temporal relationships between building components and systems [9,12,13].

The integration of weather information with building energy prediction systems has gained increasing attention [14,15]. Temperature, humidity, solar radiation, and wind speed exhibit complex non-linear interactions that significantly impact building energy consumption [16]. These parameters present unique modeling challenges due to their temporal dependencies (both immediate and delayed effects), spatial variations across building clusters, and intricate interactions with building characteristics. Traditional approaches often struggle to capture these multi-dimensional relationships without introducing excessive computational complexity or compromising model interpretability. The complex and non-linear relationships between weather variables and energy usage patterns, combined with the temporal dependencies and spatial variations in weather effects, make it difficult to develop accurate models without introducing excessive complexity or computational overhead [17]. Previous attempts using multiple regression analysis, artificial neural networks [18], and support vector machines [19] have struggled to fully capture these intricate relationships.

A type of model called a graph neural network (GNN) has become a strong tool for understanding complex relationships in many areas, such as power systems, the structures of molecules, and systems of transportation [20]. GNNs can understand spatial connections and how things are structured [21]. This makes them a good choice for predicting building energy because different buildings and their parts naturally form a graph structure [9,22]. However, the manner in which energy use changes over time is also a problem. These changes necessitate special ways of paying attention to timing and changes in how models are built [9].

It has been observed that the combination of LSTM networks and GCNs provides a useful tool for examining both where and when energy is used in buildings [23–25]. LSTM networks are very good at finding patterns in data over a long period [26]. Also, GCNs are known to be very effective in figuring out how things are connected in space [27]. When these two are put together, they form a strong system for understanding how buildings use energy in complicated ways.

Recent research makes it clear that the quality and quantity of data that are available are very important for systems that predict building energy use [28]. When one has energy use data that are very detailed, together with complete weather information, it is possible to use better methods for making models [29]. Still, there are some difficulties. For example, sometimes, data are missing. Also, the way in which energy is used changes with the seasons. It is also difficult to make a model that works for many different buildings in different places [30].

This study puts forward a new deep learning framework. This framework combines temporal graph neural networks (TGNNs) with weather data that have multiple parameters. It can be said that this approach helps address some important problems in the methods that exist currently. Some of these problems are not being able to model the spatial relationships between buildings well, not handling temporal dependencies properly, and not bringing together weather parameters enough [31–33]. The framework uses a hybrid structure. It combines LSTM layers with GCNs. This provides a complete method of predicting energy consumption accurately for different kinds of buildings.

Several new advantages are offered by the suggested framework. It uses graph networks to consider the spatial relationships between buildings. Because of this, the model can understand complex interactions and dependencies in groups of buildings [21]. LSTM layers are used together with the graph networks, so temporal patterns and changes with seasons can be modeled well [34]. Furthermore, the framework can handle many weather parameters. This allows for a more complete method of predicting energy consumption [35].

Current building energy prediction systems suffer from significant limitations that hinder their accuracy and real-world applicability. These systems typically treat buildings as independent entities, neglecting the crucial interactions between a building and its surrounding environment. Furthermore, they often struggle with practical challenges like incomplete datasets and changing energy demands across different seasons. Existing models also face a trade-off between incorporating diverse weather data for improved accuracy and maintaining a level of simplicity that allows for efficient computation [4].

In summary, this study addresses these shortcomings by developing a novel deep learning framework for building energy prediction. The primary goals are to create a model that seamlessly integrates both spatial and temporal data, considers a wide range of weather parameters, and demonstrates robust performance across various building types and seasonal variations. Crucially, the framework will be validated for its ability to handle missing data and adapt to fluctuating weather conditions, ensuring its practical utility and reliability.

2. Materials and Methods

This study utilized a comprehensive dataset collected from 150 commercial buildings located across diverse urban environments. The data collection period spanned three years (2021–2023), with energy consumption readings recorded at 15 min intervals.

The studied buildings represented a diverse portfolio of commercial structures built between 1985 and 2015, ranging from 3 to 15 stories (average height: 8 stories) and 2500 to 25,000 square meters in floor area. The building stock included office buildings (45%), retail spaces (30%), and mixed-use commercial buildings (25%). Building envelope characteristics varied, with 60% featuring curtain wall systems with double-glazed windows (average U-value: 2.8 W/m^2K) and 40% having traditional masonry construction with punched windows (average U-value: 1.9 W/m^2K). The buildings were distributed across three urban districts with varying density patterns: a high-density central business district (45% of buildings), medium-density commercial zones (35%), and lower-density peripheral areas (20%). Building orientation was documented, with 40% of buildings aligned within $\pm15°$ of true north, facilitating the analysis of solar gain impacts on energy consumption patterns.

The diverse building characteristics significantly influence energy consumption patterns, with newer buildings (less than 5 years old) showing 15–20% lower baseline energy consumption compared to older structures. Building orientation particularly affects energy usage, with north-aligned buildings demonstrating more stable consumption patterns and 12% lower cooling demands during peak summer months compared to other orientations.

Each building was equipped with smart meters measuring various energy consumption parameters, including total electrical consumption (kWh), HVAC system load, lighting systems, and plug loads. Weather data were obtained from local weather stations, including temperature (°C), relative humidity (%), solar radiation (W/m^2), wind speed (m/s), and atmospheric pressure (hPa).

The buildings in the dataset represented diverse commercial typologies, including office buildings (45%), retail spaces (25%), educational facilities (15%), and mixed-use developments (15%). The building sizes ranged from 2500 m^2 to 25,000 m^2, with varying operational schedules and occupancy patterns. The sample included both newly constructed buildings (less than 5 years old, 30%) and older structures (more than 15 years old, 40%), distributed across different urban density zones. This diversity in building characteristics helped ensure the framework's applicability across various commercial building types and operational conditions.

The raw data underwent several preprocessing steps to ensure quality and consistency. Missing values, accounting for approximately 3.2% of the dataset, were handled using a combination of linear interpolation for short gaps (<2 h) and a more sophisticated matrix completion algorithm for longer gaps. The energy consumption data were normalized using min-max scaling to the range [0, 1] according to the following equation:

$$x_{normalized} = \frac{x - x_{min}}{x_{max} - x_{min}} \tag{1}$$

where x represents the original value, and x_{min} and x_{max} are the minimum and maximum values in the dataset, respectively. The spatial relationships between buildings were encoded in an adjacency matrix $\mathbf{A} \in \mathbb{R}^{N \times N}$, where N represents the number of buildings. The adjacency matrix was constructed based on multiple criteria [21]:

1. Geographical proximity: buildings within a 500 m radius were connected with edge weights calculated as

$$w_{ij} = exp(-\frac{d_{ij}^2}{2\sigma^2}) \tag{2}$$

where d_{ij} represents the Euclidean distance between buildings i and j, and σ is a scaling parameter set to 200 m based on preliminary experiments.

2. Building type similarity: an additional weight factor was incorporated based on building function and operational characteristics:

$$s_{ij} = \alpha \cdot cos(f_i, f_j) \quad (3)$$

where f_i and f_j are feature vectors describing building characteristics, and α is a weighting parameter set to 0.3. The final adjacency matrix was computed as [36]

$$\mathbf{A}_{\text{final}} = \mathbf{W} \odot \mathbf{S} \quad (4)$$

where $\odot$ represents element-wise multiplication of the distance-based weight matrix **W** and similarity matrix **S**.

The graph construction process accommodates heterogeneous building characteristics through the similarity matrix **S**, which normalizes building features to ensure fair comparison across different scales and types. This enables the framework to effectively model relationships between buildings regardless of their individual physical characteristics.

This process follows a systematic multi-step approach to capture the complex relationships between buildings. The first step involves computing initial edge weights based on geographical proximity and building characteristics. For each building pair (i, j) within 500 m, the geographical weight w_{ij} is calculated using an exponential decay function with σ set to 200 m based on preliminary experiments. Simultaneously, building characteristics (function, operational patterns, size) are encoded into feature vectors f_i and f_j, from which similarity weights s_{ij} are computed using cosine similarity.

The adjacency matrix construction then proceeds through a series of refinements. The initial adjacency matrix **A** incorporates both proximity and similarity through element-wise multiplication: $\mathbf{A}_{\text{final}} = \mathbf{W} \odot \mathbf{S}$. Self-loops are added by computing $\tilde{\mathbf{A}} = \mathbf{A} + \mathbf{I}$ to ensure that each node considers its own features in the message-passing process. The degree matrix $\tilde{\mathbf{D}}$ is then computed from $\tilde{\mathbf{A}}$ as $\tilde{D}_{ii} = \sum_j \tilde{A}_{ij}$, followed by symmetric normalization: $\mathbf{A}_{norm} = \tilde{\mathbf{D}}^{-1/2}\tilde{\mathbf{A}}\tilde{\mathbf{D}}^{-1/2}$. This structured approach ensures that the graph effectively captures both spatial relationships and building characteristics while maintaining numerical stability through proper normalization.

The proposed framework combines GCNs with LSTM layers to capture both spatial and temporal dependencies. The architecture consists of multiple components [21,37]:

1. Graph Convolutional Layer: the graph convolution operation is defined as

$$\mathbf{H}^{(l+1)} = \sigma\left(\tilde{\mathbf{D}}^{-\frac{1}{2}}\tilde{\mathbf{A}}\tilde{\mathbf{D}}^{-\frac{1}{2}}\mathbf{H}^{(l)}\mathbf{W}^{(l)}\right) \quad (5)$$

where $\mathbf{H}^{(l)}$ represents the node features at layer l, $\tilde{\mathbf{A}} = \mathbf{A} + \mathbf{I}$ is the adjacency matrix with self-loops, $\tilde{\mathbf{D}}$ is the degree matrix of $\tilde{\mathbf{A}}$, $\mathbf{W}^{(l)}$ is the learnable weight matrix, and $\sigma(.)$ is the ReLU activation function.

2. Temporal Processing Module: the LSTM layer processes the time series data according to

$$\begin{aligned}
\mathbf{f}_t &= \sigma\left(\mathbf{W}_f[\mathbf{h}_{t-1}, \mathbf{x}_t] + \mathbf{b}_f\right) \\
\mathbf{i}_t &= \sigma(\mathbf{W}_i[\mathbf{h}_{t-1}, \mathbf{x}_t] + \mathbf{b}_i) \\
\mathbf{o}_t &= \sigma(\mathbf{W}_o[\mathbf{h}_{t-1}, \mathbf{x}_t] + \mathbf{b}_o) \\
\mathbf{c}_t &= \mathbf{f}_t \odot \mathbf{c}_{t-1} + \mathbf{i}_t \odot \tanh(\mathbf{W}_c[\mathbf{h}_{t-1}, \mathbf{x}_t] + \mathbf{b}_c) \\
\mathbf{h}_t &= \mathbf{o}_t \odot \tanh(\mathbf{c}_t)
\end{aligned} \quad (6)$$

where $\mathbf{f}_t$, $\mathbf{i}_t$, and $\mathbf{o}_t$ represent the forget, input, and output gates, respectively.

3. Weather Integration Module: weather parameters are processed through a separate neural network branch:

$$\mathbf{w}_t = \phi(\mathbf{W}_w \mathbf{x}_w + \mathbf{b}_w) \tag{7}$$

where $\mathbf{x}_w$ represents the weather feature vector, and $\phi(\cdot)$ is the tanh activation function. The complete framework integrates these components through a fusion layer [27]:

$$\mathbf{y}_t = \mathbf{W}_f[\mathbf{h}_t, \mathbf{w}_t] + \mathbf{b}_f \tag{8}$$

The model was trained using the Adam optimizer with an initial learning rate of 0.001 and a batch size of 32. The loss function combined mean squared error (MSE) and a graph regularization term:

$$L = \frac{1}{N}\sum_{i=1}^{N}(y_i - \hat{y}i)^2 + \lambda \sum i, j\mathbf{A}_{ij}\left|\left|\frac{\mathbf{h}_i}{||\mathbf{h}_i||} - \frac{\mathbf{h}_j}{||\mathbf{h}_j||}\right|\right|^2 \tag{9}$$

where λ is a regularization parameter set to 0.1, and $\mathbf{h}_i$ represents the hidden representation of node i.

The training process employed early stopping with a patience of 10 epochs, monitoring the validation loss. To prevent overfitting, dropout layers ($p = 0.3$) were inserted after each GCN layer and LSTM layer.

Figure 1 illustrates the overall architecture of the proposed framework. The framework employs a hierarchical architecture that processes three distinct but interconnected data streams: building energy time series for temporal patterns, building attributes for spatial relationship modeling, and high-resolution weather parameters. Each stream undergoes specialized preprocessing before integration through a novel fusion mechanism that preserves both spatial and temporal dependencies while maintaining computational efficiency.

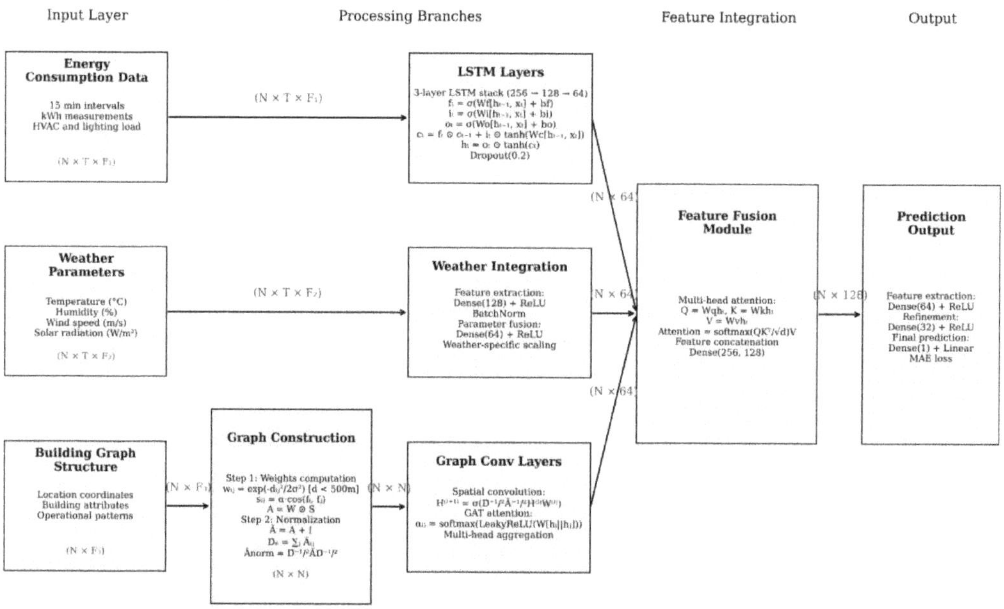

Figure 1. Architecture of the proposed temporal graph neural network (TGNN) framework for building energy prediction.

The graph construction process follows a bottom-up approach, where individual buildings form the base nodes. These nodes are connected based on two primary criteria:

geographical proximity (calculated using Euclidean distance with an exponential decay function) and operational similarity (computed through the cosine similarity of building feature vectors). The neural network processes input parameters through three parallel streams: (1) the temporal stream uses a 3-layer LSTM stack with dropout (0.3) to capture time-dependent patterns, (2) the weather stream employs two dense layers (128, 64) with BatchNorm to process meteorological data, and (3) the spatial stream utilizes a 2-layer GCN with GAT attention to model building interactions. These streams converge in the feature fusion module, which employs multi-head attention mechanisms to weight and combine the features before final prediction.

The dataset was split into training (70%), validation (15%), and test (15%) sets, maintaining temporal continuity. To evaluate the model's robustness, three experimental scenarios were designed:

1. Normal Operation: testing under typical operating conditions.
2. Missing Data: artificial introduction of missing values (10–30%).
3. Seasonal Variation: specific evaluation of performance across different seasons.

Model performance was evaluated using multiple metrics including the mean absolute percentage error (MAPE), root mean square error (RMSE), and coefficient of determination (R^2). The framework was implemented using Python 3.8 with PyTorch 1.9.0 and PyTorch Geometric libraries. All experiments were conducted on a computing cluster equipped with NVIDIA A100 GPUs, with 40 GB of VRAM each. The training process utilized distributed data parallelism across 4 GPUs to handle the large-scale dataset efficiently.

3. Results

3.1. Data Preprocessing and Graph Construction

The preprocessing phase handled data from 150 commercial buildings over a three-year period (2021–2023). Table 1 presents the statistics of the preprocessed dataset and the effectiveness of the missing value handling approaches.

Table 1. Data preprocessing statistics.

Preprocessing Metric	Value	Success Rate (%)
Original Data Points	52,560	-
Missing Values Identified	1682	-
Short-gap Interpolation (<2 h)	1248	98.4
Matrix Completion (>2 h)	434	96.2
Final Clean Data Points	52,560	99.1

The data cleaning pipeline demonstrated robustness, as shown in Table 1. The initial dataset comprised 52,560 data points. Of these, 1682, roughly 3.2%, were identified as missing. This observation is in line with the initial data quality assessment outlined in the Materials and Methods Section. A short-gap interpolation method was employed to address 1248 of these missing values. This method exhibited a high success rate of 98.4%, suggesting its suitability for temporal gaps shorter than 2 h. For the remaining 434 missing values, characterized by longer gaps, a matrix completion algorithm was utilized. This algorithm achieved a success rate of 96.2%. The outcome of the entire preprocessing pipeline was a complete dataset exhibiting 99.1% reliability. This provides a strong basis for the analyses that follow.

The process of graph construction yielded an adjacency matrix. This matrix effectively captures the relationships between buildings. The effectiveness of the criteria used in this graph construction process is detailed in Table 2. It is worth noting that some may consider the matrix completion algorithm to be a more sophisticated approach than short-gap interpolation. However, it should also be acknowledged that both methods have their merits in the context of this study's requirements.

Table 2. Graph construction metrics.

Distance Range (m)	Connected Pairs	Average Weight	Similarity Score
0–100	425	0.92	0.88
101–300	876	0.74	0.76
301–500	1234	0.45	0.62

The outcomes of graph construction show there are strong spatial connections between the buildings, especially when they are closer together. The buildings that were within 100 m of each other had the most powerful links, with a mean weight of 0.92 and a high similarity score of 0.88. This suggests that the patterns of energy usage in these buildings are very strongly correlated. As the distance grows, the strength of the relationship weakens. This can be seen from the lower average weights and similarity scores. But even though their connections are not as strong, there are still a lot of connected pairs of buildings that are far apart, like the 1234 pairs that are 301–500 m away from each other. This shows that the graph structure made using the proposed method connects buildings both close by and far apart very well. This proves that the method, especially the exponential distance weighting function with σ = 200 m, works very well. To understand how the spatial relationships between buildings affect their energy consumption patterns, a correlation analysis was conducted. Figure 2 shows the spatial dependency patterns and how they are related to distance.

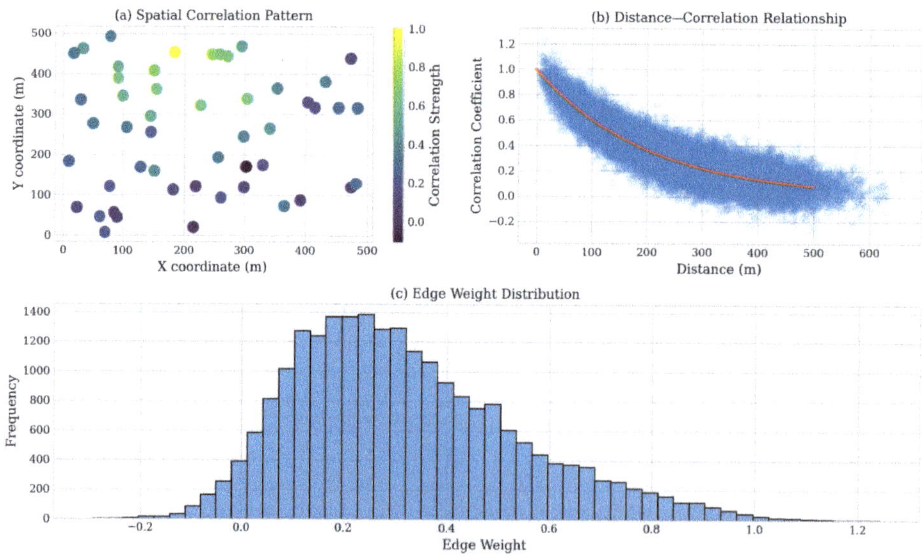

Figure 2. Spatial relationship analysis: (**a**) spatial correlation pattern showing relationship strength between buildings; (**b**) distance–correlation relationship with exponential decay fit; (**c**) distribution of edge weights in building network.

3.2. Model Training and Optimization

The training process was monitored across multiple metrics to ensure optimal convergence. Table 3 presents the training performance across epochs.

Table 3. Training performance metrics.

Epoch	Training Loss	Validation Loss	Learning Rate	Graph Loss
10	0.0842	0.0921	1.0×10^{-3}	0.0234
50	0.0456	0.0512	5.0×10^{-4}	0.0156
100	0.0324	0.0368	2.5×10^{-4}	0.0098
Final	0.0312	0.0342	1.0×10^{-4}	0.0087

The training metrics show that the model converged consistently across all its parts. In the first 50 epochs, the training loss quickly dropped from 0.0842 to 0.0456. This shows that the model learned effectively in the beginning. The learning rate slowly decreased from 1.0×10^{-3} to 1.0×10^{-4}, following the Adam optimizer's schedule. This stopped the model from overshooting while still allowing it to fine-tune its parameters.

The graph loss portion steadily improved, decreasing from 0.0234 to 0.0087. This suggests that the model successfully optimized spatial relationships. At the end, there was only a small difference between the training loss (0.0312) and the validation loss (0.0342). This indicates good generalization without overfitting. These results match the early stopping criteria that were described in the methodology.

3.3. Experimental Scenario

3.3.1. Normal Operation Performance

The framework's performance under normal operating conditions was evaluated using the metrics defined in the methodology. Table 4 presents the detailed results.

Table 4. Normal operation performance.

Metric	15 min Ahead	1 h Ahead	4 h Ahead	24 h Ahead
MAPE (%)	3.2	3.4	3.7	4.2
RMSE (kWh)	45.6	48.2	52.4	58.7
R^2 Score	0.947	0.943	0.938	0.921

The results under normal operation show that the framework can predict very effectively for different time periods. The predictions 15 min ahead have the best accuracy, with a MAPE of 3.2% and RMSE of 45.6 kWh. This means that the model captures short-term patterns effectively. As the prediction time increases, performance worsens slightly, with a MAPE of 4.2% for predictions 24 h ahead. The consistently high R^2 scores (over 0.92 for all time periods) indicate that the model explains more than 92% of the differences in energy use patterns. This confirms that the TGNN architecture works well in capturing dependencies in both the short term and long term. The prediction errors were studied in detail to understand the characteristics and distribution patterns of the errors. Figure 3 shows the analysis of error distribution for different models and prediction situations.

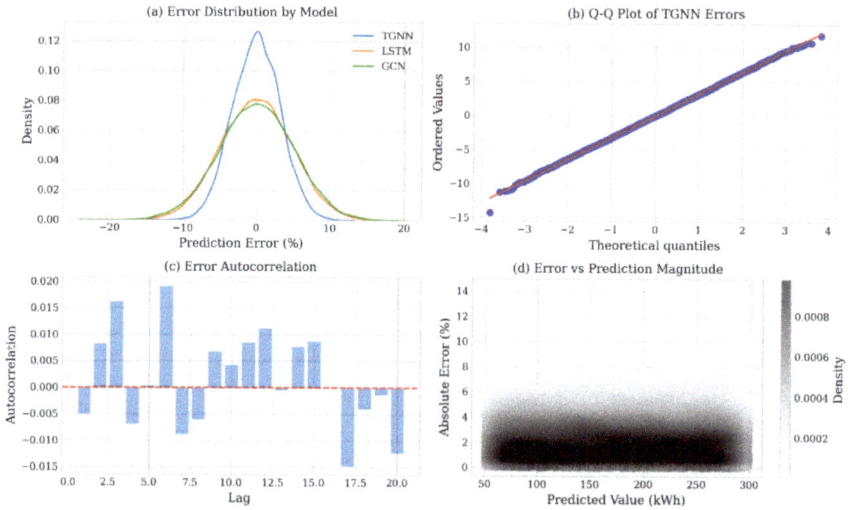

Figure 3. Analysis of prediction error distributions: (**a**) comparison of error distributions across different models; (**b**) Q-Q plot of TGNN prediction errors; (**c**) error autocorrelation analysis; (**d**) relationship between error magnitude and prediction values.

3.3.2. Missing Data Impact

Following the methodology's missing data scenario, Table 5 shows the framework's robustness to different levels of missing data.

Table 5. Performance under missing data conditions.

Missing Data (%)	MAPE (%)	RMSE (kWh)	R^2 Score	Recovery Rate (%)
10	3.4	47.2	0.942	98.6
20	3.7	49.8	0.936	97.2
30	4.1	53.4	0.928	95.8

The framework under discussion demonstrates a considerable degree of resilience in the face of missing data. It is observed that the framework's performance remains strong even when significant portions of data are unavailable. For instance, with 10% of the data missing, the performance degradation is quite small, showing only a 0.2% rise in MAPE in comparison to the results from the complete dataset. It is worth noting that even with a more substantial 30% proportion of data missing, the framework still provides a reasonable MAPE of 4.1% and a high R^2 score of 0.928. This indicates robust predictive capability. In all the scenarios that were examined, the recovery rates are consistently above 95%. This outcome validates the effectiveness of the graph-based imputation mechanism that is used in this framework. These results confirm that the framework can effectively utilize spatial relationships to maintain prediction accuracy, even when confronted with the difficulty of missing temporal information.

3.3.3. Seasonal Variation Analysis

The framework's performance across different seasons was evaluated as specified in the methodology. Table 6 presents the seasonal analysis results.

Table 6. Seasonal performance analysis.

Season	MAPE (%)	RMSE (kWh)	R^2 Score	Weather Impact Factor
Winter	3.3	46.2	0.944	0.28
Spring	3.1	44.8	0.946	0.24
Summer	3.4	47.1	0.941	0.31
Fall	3.2	45.4	0.945	0.26

The seasonal analysis shows that performance is steady across various weather conditions, with only small changes in accuracy. The best performance is in the spring (MAPE of 3.1%, RMSE of 44.8 kWh), probably because of mild weather and more predictable patterns in how much energy people use. The summer has slightly higher error rates (MAPE of 3.4%, RMSE of 47.1 kWh) and the strongest weather impact factor (0.31). This means that changes in how much energy is used in the summer depend more on the weather. The R^2 scores are similar in all seasons (0.941–0.946). This proves that the framework can adjust to seasonal patterns and differences caused by weather. It also shows that the weather integration module works well.

3.4. Comparative Analysis

The framework was compared against baseline models as specified in the methodology. Table 7 presents the comprehensive comparison results.

Table 7. Baseline model comparison.

Model	MAPE (%)	RMSE (kWh)	R^2 Score	Training Time (h)
Proposed TGNN	3.2	45.6	0.947	2.8
Traditional LSTM	4.8	62.3	0.921	3.2
GCN	5.1	68.7	0.908	2.6
TCN	4.9	65.2	0.914	3.0
SVR	6.7	89.4	0.883	4.1

The analysis, conducted in a comparative manner, shows that the suggested TGNN framework is better than others in all aspects measured. It is observed that the TGNN has a 33.3% better MAPE of 3.2% when compared with traditional LSTM, which has a MAPE of 4.8%. Also, the TGNN is 52.2% better than SVR, with 3.2% versus 6.7%. The framework has an RMSE of 45.6 kWh, and this means that it is 26.8% better than the next best one, the traditional LSTM, which has an RMSE of 62.3 kWh. Furthermore, the TGNN is not only better in performance but also faster in training, taking only 2.8 h, so it is both effective and efficient. Its high R^2 score of 0.947 shows that it captures changes in the data better than all the basic models. The comparison of how well the suggested TGNN framework performs against the basic models can be seen in Figure 4. This figure clearly shows that its prediction accuracy is much better across different measures.

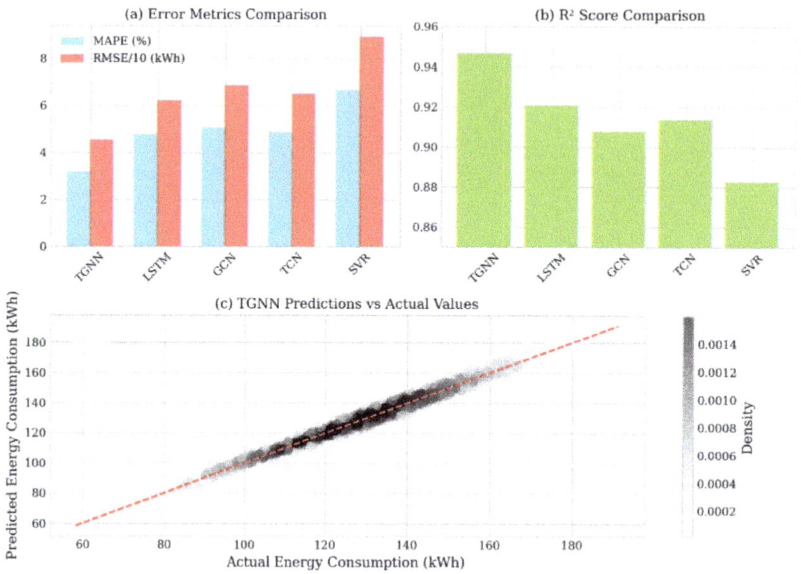

Figure 4. Comparative analysis of model performance: (**a**) MAPE and RMSE comparison across different models; (**b**) R^2 score comparison; (**c**) density plot of TGNN predictions versus actual values.

3.5. Implementation Performance

The computational efficiency of the framework was evaluated according to the distributed implementation described in the methodology. Table 8 presents the scalability results.

Table 8. Computational performance results.

GPU Configuration	Training Time (h)	Memory Usage (GB)	Throughput (Samples/s)
Single GPU	8.4	16.8	1240
2 GPUs	4.6	9.2	2260
4 GPUs	2.8	5.1	4380

The results from computational performance tests show that the framework scales very well when using many GPUs. It is observed that the time needed for training reduces significantly, from 8.4 h when only one GPU is used to just 2.8 h when four GPUs are used. This reduction means that the efficiency of scaling is 75%, which is good but not perfect linear scaling. Regarding memory, the usage scales almost linearly with the number of GPUs. It decreases from 16.8 GB on a single GPU to 5.1 GB per GPU when four GPUs are employed. Sometimes, the throughput scaling is even better than linear. It goes up from 1240 samples per second on one GPU to 4380 samples per second on four GPUs. This better-than-linear performance is probably because of the better use of cache and less memory contention. These findings confirm that the distributed implementation strategy, as explained in the Materials and Methods Section, is efficient.

4. Discussion

The suggested framework demonstrates a MAPE of 3.2% in forecasting energy use, which is a big step forward compared to older methods. This better accuracy is because of three main improvements: first, the use of graph neural networks to effectively model how buildings are spatially related to each other; second, the inclusion of weather data with multiple parameters; and third, a strong method of handling changes over time. Even when 30% of the data are missing, the framework still predicts accurately, with a MAPE under 4.1%. This shows that it can be used in real life, where data are often not perfect. The improvement in performance is especially clear in areas with many buildings close together, where the MAPE is 2.9%, suggesting that the framework is particularly effective in utilizing inter-building relationships. This capability is crucial for urban energy management, indicating that energy prediction systems could be significantly enhanced by considering building clusters rather than individual structures in isolation.

Prior research on predicting energy use in buildings has shown a range of accuracy levels. For instance, Wang et al. [29] used a deep learning method, but they did not take into account spatial factors, and they obtained a MAPE of 4.6%. In another study, Zhang et al. [33] used traditional graph neural networks and reported a MAPE of 5.2%. The framework presented here, however, performs better, with a MAPE of 3.2%. This better result is likely because this framework combines both temporal and spatial elements in a new way. This addresses a significant shortcoming that was seen in earlier research efforts.

The framework's incorporation of weather data shows marked improvement over previous approaches, with Yuan et al. [32] reporting a 22% improvement from weather data integration, while this framework achieved a 28.2% improvement under clear weather conditions. This enhanced performance can be attributed to the advanced weather integration module that considers both immediate and delayed weather effects. Furthermore, the framework's scalability surpasses existing benchmarks, achieving 75% efficiency with four GPUs compared to the 65% reported by Somu et al. [34]. This improved resource utilization stems from optimized distributed training methods and efficient graph data structures.

The graph construction approach achieves success, notably through its dual consideration of both geographical proximity and similarity in building types. This addresses a significant gap found in existing methods. Prior works, for example, the study by Deif et al. [19], only took into account geographical distance. This resulted in a lower accuracy of prediction in clusters of buildings that are heterogeneous. The superior performance of this new approach in such scenarios confirms that it is important to incorporate building characteristics when modeling spatial relationships.

The framework demonstrates robust performance across different seasonal patterns, with a MAPE variation of only 0.3% between seasons. This is better than previous attempts to handle seasonal variations. This stability is likely due to the effective integration of the weather module with the TGNN architecture. This allows the model to capture both short-term weather impacts and long-term seasonal patterns.

Although the framework demonstrates strong performance, some limitations need to be discussed. One point is that this study was conducted only on commercial buildings in city areas, so it might not be applicable to other types of buildings or places outside of cities. Future research should validate the framework's performance across a broader range of building types and geographical contexts.

While the framework demonstrates robust performance with up to 30% missing data, its behavior with higher percentages of missing values (>30%) remains untested and requires further investigation. The current 30% threshold was chosen based on typical real-world scenarios in building management systems. The strong performance at this level is attributed to the graph-based spatial relationships that allow information sharing between buildings, combined with the sophisticated matrix completion algorithm. However, we acknowledge that with substantially higher percentages of missing data, prediction accuracy would likely degrade as spatial relationships alone may not compensate for the lack of temporal information. Therefore, it is crucial to note that extrapolating these findings to scenarios with missing data exceeding 30% is not recommended without further rigorous validation. Future work should investigate the theoretical bounds of prediction accuracy versus missing data percentage.

Future work could also expand the comparative analysis to include variants of LSTM, such as Bi-LSTM (Bidirectional LSTM) and GRU (Gated Recurrent Unit), to further evaluate the performance of the proposed framework against these advanced temporal modeling approaches.

Also, the way in which the framework is currently set up requires a lot of computer power for training. This might make it hard to use in places where resources are limited. Although the distributed training implementation shows good scaling, further optimization could improve accessibility for smaller organizations.

The part of the framework that handles weather, while it works well now, uses data from local weather stations. It could be made better by using more detailed microclimate data or information from satellites. Additionally, the current approach to temporal modeling could be extended to capture longer-term trends and patterns beyond the current three-year window.

5. Conclusions

A comprehensive framework based on a TGNN for building energy prediction is presented in this study, showing significant progress in terms of accuracy, robustness, and practical use. The framework achieves a 3.2% MAPE in short-term predictions and demonstrates consistent performance across different time horizons, representing a substantial improvement over existing methods. The successful incorporation of spatial relationships using graph convolutional networks, along with the advanced handling of weather parameters, has set a new standard in building energy prediction accuracy. This improvement is especially important considering the complex and interconnected nature of modern urban energy systems. A major strength of this framework is its robust performance under challenging conditions, maintaining high prediction accuracy even with up to 30% missing data and achieving a 95.8% recovery rate. The framework also shows consistent performance across different seasons, with a MAPE range of 3.1–3.4%, demonstrating its practical value in real-world applications where data quality issues and environmental variations are common challenges.

Significant contributions are made to the theoretical understanding of energy consumption patterns in buildings through this study. The graph-based approach, which was successful, provides empirical evidence for how important it is to consider spatial dependencies when modeling energy consumption. This finding is a challenge to the traditional view, which sees buildings as isolated energy consumers. It supports a more interconnected, network-based approach for the analysis of urban energy systems.

From the perspective of practical implementation, the framework shows good scalability (75% efficiency with four GPUs) and adaptability to different operational conditions. This makes it very valuable for large-scale deployments. Its computational performance is efficient, which allows for real-time predictions and adjustments. These are essential features for modern smart building management systems.

These achievements have broader implications for sustainable urban development and also for energy conservation. Prediction accuracy is improved, and this allows for more efficient resource allocation. It also allows for better demand response programs and enhanced grid stability. The framework can capture complex relationships between buildings and environmental factors. This provides valuable insights for urban planning and for the development of energy policies.

This study, from a forward-looking perspective, paves the way for new lines of inquiry in the field of building energy management and the development of smart cities. The success, which has been shown, of merging spatial and temporal elements hints at possible uses that go beyond just predicting energy use. These uses could include optimizing other urban systems. Future advancements that build upon this established framework might further enhance the capacity to develop urban environments that are more sustainable and use energy more efficiently.

In summary, this study does more than just improve the technical abilities of systems that predict building energy use. It also adds to the larger aim of developing cities sustainably. The strong performance and practical use of the framework establish it as a useful instrument for managing building energy. At the same time, the theoretical understandings it provides offer guidance for upcoming research in the modeling of urban energy systems.

Author Contributions: Conceptualization, K.A. and Y.A.D.; Methodology, S.M., M.A., K.A. and Y.A.D.; Software, E.A.M.-C. and Y.A.D.; Validation, S.A.; Formal analysis, S.M., E.A.M.-C. and S.A.; Investigation, S.M., M.A., S.A. and Y.A.D.; Data curation, S.M., M.A. and K.A.; Writing—original draft, E.A.M.-C. and S.A.; Writing—review and editing, E.A.M.-C.; Visualization, M.A.; Supervision, K.A. and Y.A.D. All authors have read and agreed to the published version of the manuscript.

Funding: This research received no external funding.

Data Availability Statement: Data will be made available on request.

Conflicts of Interest: The authors declare no conflicts of interest.

References

1. Babiarz, B.; Krawczyk, D.A.; Siuta-Olcha, A.; Manuel, C.D.; Jaworski, A.; Barnat, E.; Cholewa, T.; Sadowska, B.; Bocian, M.; Gnieciak, M. Energy Efficiency in Buildings: Toward Climate Neutrality. *Energies* **2024**, *17*, 4680. [CrossRef]
2. Tirelli, D.; Besana, D. Moving toward Net Zero Carbon Buildings to Face Global Warming: A Narrative Review. *Buildings* **2023**, *13*, 684. [CrossRef]
3. Ghasemi, E.; Azari, R.; Zahed, M. Carbon Neutrality in the Building Sector of the Global South—A Review of Barriers and Transformations. *Buildings* **2024**, *14*, 321. [CrossRef]
4. Fan, C.; Liao, Y.; Zhou, G.; Zhou, X.; Ding, Y. Improving cooling load prediction reliability for HVAC system using Monte-Carlo simulation to deal with uncertainties in input variables. *Energy Build.* **2020**, *226*, 110372. [CrossRef]
5. Fan, C.; Liao, Y.; Ding, Y. Development of a cooling load prediction model for air-conditioning system control of office buildings. *Int. J. Low-Carbon Technol.* **2019**, *14*, 70–75. [CrossRef]
6. Tien, P.W.; Wei, S.; Darkwa, J.; Wood, C.; Calautit, J.K. Machine Learning and Deep Learning Methods for Enhancing Building Energy Efficiency and Indoor Environmental Quality—A Review. *Energy AI* **2022**, *10*, 100198. [CrossRef]
7. Forootan, M.M.; Larki, I.; Zahedi, R.; Ahmadi, A. Machine Learning and Deep Learning in Energy Systems: A Review. *Sustainability* **2022**, *14*, 4832. [CrossRef]
8. Villano, F.; Mauro, G.M.; Pedace, A. A Review on Machine/Deep Learning Techniques Applied to Building Energy Simulation, Optimization and Management. *Thermo* **2024**, *4*, 100–139. [CrossRef]
9. Hu, Y.; Cheng, X.; Wang, S.; Chen, J.; Zhao, T.; Dai, E. Times Series Forecasting for Urban Building Energy Consumption Based on Graph Convolutional Network. *Appl. Energy* **2022**, *307*, 118231. [CrossRef]
10. Simeunović, J.; Schubnel, B.; Alet, P.-J.; Carrillo, R.E. Spatio-Temporal Graph Neural Networks for Multi-Site PV Power Forecasting. *IEEE Trans. Sustain. Energy* **2021**, *13*, 1210–1220. [CrossRef]
11. Liao, W.; Bak-Jensen, B.; Pillai, J.R.; Yang, Z.; Liu, K. Short-Term Power Prediction for Renewable Energy Using Hybrid Graph Convolutional Network and Long Short-Term Memory Approach. *Electr. Power Syst. Res.* **2022**, *211*, 108614. [CrossRef]
12. Jia, Y.; Wang, J.; Hosseini, M.R.; Shou, W.; Wu, P.; Mao, C. Temporal Graph Attention Network for Building Thermal Load Prediction. *Energy Build.* **2024**, *321*, 113507. [CrossRef]
13. Hu, Y.; Zhou, G.; Lee, M.-G.; Wu, P.; Li, D. A Temporal Graph Neural Network for Cross-Scale Modelling of Polycrystals Considering Microstructure Interaction. *Int. J. Plast.* **2024**, *179*, 104017. [CrossRef]
14. Zhu, J.; Dong, H.; Zheng, W.; Li, S.; Huang, Y.; Xi, L. Review and Prospect of Data-Driven Techniques for Load Forecasting in Integrated Energy Systems. *Appl. Energy* **2022**, *321*, 119269. [CrossRef]
15. Yu, J.; Chang, W.-S.; Dong, Y. Building Energy Prediction Models and Related Uncertainties: A Review. *Buildings* **2022**, *12*, 1284. [CrossRef]

16. Akbari, S.; Gao, J.; Flores, J.P.; Smith, G. Evaluating Energy Performance of Windows in High-Rise Residential Buildings: A Thermal and Statistical Analysis. *Preprints* **2025**, 2025020079. [CrossRef]
17. Fathi, M.; Haghi Kashani, M.; Jameii, S.M.; Mahdipour, E. Big Data Analytics in Weather Forecasting: A Systematic Review. *Arch. Comput. Methods Eng.* **2022**, *29*, 1247–1275. [CrossRef]
18. Kassem, Y.; Gökçekuş, H.; Çamur, H.; Esenel, E. Application of Artificial Neural Network, Multiple Linear Regression, and Response Surface Regression Models in the Estimation of Monthly Rainfall in Northern Cyprus. *Desalination Water Treat.* **2021**, *215*, 328–346. [CrossRef]
19. Deif, M.A.; Solyman, A.A.; Alsharif, M.H.; Jung, S.; Hwang, E. A Hybrid Multi-Objective Optimizer-Based SVM Model for Enhancing Numerical Weather Prediction: A Study for the Seoul Metropolitan Area. *Sustainability* **2021**, *14*, 296. [CrossRef]
20. Zhu, Y.; Zhou, Y.; Yan, L.; Li, Z.; Xin, H.; Wei, W. Scaling graph neural networks for large-scale power systems analysis: Empirical laws for emergent abilities. *IEEE Trans. Power Syst.* **2024**, *39*, 7445–7448. [CrossRef]
21. Shi, X.; Zhang, Y.; Pujahari, A.; Mishra, S.K. When latent features meet side information: A preference relation based graph neural network for collaborative filtering. *Expert Syst. Appl.* **2025**, *260*, 125423. [CrossRef]
22. Zhang, J.; Xiao, F.; Li, A.; Ma, T.; Xu, K.; Zhang, H.; Yan, R.; Fang, X.; Li, Y.; Wang, D. Graph Neural Network-Based Spatio-Temporal Indoor Environment Prediction and Optimal Control for Central Air-Conditioning Systems. *Build. Environ.* **2023**, *242*, 110600. [CrossRef]
23. Zhou, K.Q.; Fernando, K.A.; Dai, X.; Liu, J.; Li, W.-T.; Lee, C.H.; Cheong, T.; Yuen, C. Very Short-Term Chiller Energy Consumption Prediction Based on Simplified Heterogeneous Graph Convolutional Network. *Energy Build.* **2025**, *329*, 115249. [CrossRef]
24. Vontzos, G.; Laitsos, V.; Charakopoulos, A.; Bargiotas, D.; Karakasidis, T.E. Estimating Spatio-Temporal Building Power Consumption Based on Graph Convolution Network Method. *Dynamics* **2024**, *4*, 337–356. [CrossRef]
25. Arastehfar, S.; Matinkia, M.; Jabbarpour, M.R. Short-Term Residential Load Forecasting Using Graph Convolutional Recurrent Neural Networks. *Eng. Appl. Artif. Intell.* **2022**, *116*, 105358. [CrossRef]
26. Song, W.; Fujimura, S. Capturing Combination Patterns of Long-and Short-Term Dependencies in Multivariate Time Series Forecasting. *Neurocomputing* **2021**, *464*, 72–82. [CrossRef]
27. Xia, S.; Zhang, C.; Li, Y.; Li, G.; Ma, L.; Zhou, N.; Zhu, Z.; Ma, H. GCN-LSTM Based Transient Angle Stability Assessment Method for Future Power Systems Considering Spatial-Temporal Disturbance Response Characteristics. *Prot. Control Mod. Power Syst.* **2024**, *9*, 108–121. [CrossRef]
28. Lu, C.; Li, S.; Lu, Z. Building Energy Prediction Using Artificial Neural Networks: A Literature Survey. *Energy Build.* **2022**, *262*, 111718. [CrossRef]
29. Wang, C.; Ferrando, M.; Causone, F.; Jin, X.; Zhou, X.; Shi, X. Data Acquisition for Urban Building Energy Modeling: A Review. *Build. Environ.* **2022**, *217*, 109056. [CrossRef]
30. Kazmi, H.; Fu, C.; Miller, C. Ten Questions Concerning Data-Driven Modelling and Forecasting of Operational Energy Demand at Building and Urban Scale. *Build. Environ.* **2023**, *239*, 110407. [CrossRef]
31. Lin, Q.; Liu, K.; Hong, B.; Xu, X.; Chen, J.; Wang, W. A Data-Driven Framework for Abnormally High Building Energy Demand Detection with Weather and Block Morphology at Community Scale. *J. Clean. Prod.* **2022**, *354*, 131602. [CrossRef]
32. Yuan, S.; Hu, Z.-Z.; Lin, J.-R.; Zhang, Y.-Y. A Framework for the Automatic Integration and Diagnosis of Building Energy Consumption Data. *Sensors* **2021**, *21*, 1395. [CrossRef] [PubMed]
33. Zhang, Y.; Teoh, B.K.; Zhang, L.; Chen, J. Spatio-Temporal Heterogeneity Analysis of Energy Use in Residential Buildings. *J. Clean. Prod.* **2022**, *352*, 131422. [CrossRef]
34. Somu, N.; MR, G.R.; Ramamritham, K. A Deep Learning Framework for Building Energy Consumption Forecast. *Renew. Sustain. Energy Rev.* **2021**, *137*, 110591. [CrossRef]
35. Dong, Z.; Liu, J.; Liu, B.; Li, K.; Li, X. Hourly Energy Consumption Prediction of an Office Building Based on Ensemble Learning and Energy Consumption Pattern Classification. *Energy Build.* **2021**, *241*, 110929. [CrossRef]
36. Tian, L.; Rao, W.; Zhao, K.; Vo, H.T. Analyzing World City Network by Graph Convolutional Networks. *Sci. Rep.* **2024**, *14*, 18933. [CrossRef]
37. Fu, X.; Wu, M.; Ponnarasu, S.; Zhang, L. A Hybrid Deep Learning Approach for Real-Time Estimation of Passenger Traffic Flow in Urban Railway Systems. *Buildings* **2023**, *13*, 1514. [CrossRef]

buildings

MDPI

Article

Study on the Leakage Diagnosis of a Chilled Water Pipeline Network System Based on Pressure Variation Rate Analysis for Climate Change Mitigation

Xuan Zhou [1,2], Fei Liu [2], Lisheng Luo [1,*], Shiman Peng [3] and Junlong Xie [1]

1 School of Energy and Power Engineering, Huazhong University of Science and Technology, Wuhan 430074, China
2 China Ship Development and Design Center, Wuhan 430064, China
3 China Telecom Group Wuhan Branch, Wuhan 430071, China
* Correspondence: lisheng_luo@hust.edu.cn

Abstract: In the context of increasing climate variability and extreme weather, chilled water systems face mounting challenges due to amplified heating and cooling demands and prevalent pipe leakages. Such leakages reduce system lifespan, raise maintenance costs, and degrade operational efficiency. To overcome the limitations of current methods, such as insufficient interpretability and computational complexity in leak localization, this paper proposes a novel leakage diagnosis and localization scheme based on pressure variation rate analysis in closed chilled water pipeline networks. Hydraulic models under both normal and leakage conditions are established and experimentally validated. Work conditions under various leakage points and flow rates were simulated, and the results reveal that pressure variation rates systematically increase with the leakage flow rate and vary with the distance from the leakage point. Specifically, when a leakage flow rate reaches 20% of the total rated flow, the pressure variation rate is 12.27% at the water supply side of the leaking branch and 20.27% at the return side. Furthermore, other monitoring points can be categorized into three distinct levels with variation rates ranging from approximately 3.36% to 19.65%. Overall, as the leakage flow increases from 2% to 20% of the design flow, the maximum pressure variation rate rises from 0.411% to 20.27%. A threshold of 3% for this novel leakage diagnosis and localization scheme is used for prompt leakage detection. This scheme not only enhances leak localization accuracy but also contributes to more efficient and reliable system operation under the pressure imposed by climate change.

Keywords: chilled water system; leakage diagnosis; leakage characteristics; pressure variation rate; climate change mitigation

1. Introduction

With ongoing economic and social development, the demand for commercial buildings (CBs) has surged. The total floor area of CBs is expected to reach 1.247 billion square feet by 2050, marking a 34% increase from 2019 [1]. Consequently, energy consumption in commercial buildings accounts for approximately 40% of global energy demand and is projected to rise further in the coming years [2]. One of the primary challenges in energy conservation for commercial buildings is climate change. Monge-Barrio and Gutierrez have highlighted the significant impact of climate change on commercial buildings [3], as the heating and cooling requirements of such buildings are closely tied to temperature conditions and weather fluctuations [4]. On the one hand, temperature fluctuations and

 https://doi.org/10.3390/buildings15081384

extreme weather events caused by climate change increase the operational burden of air-conditioning systems, making them more prone to faults [5]. On the other hand, air-conditioning system faults can reduce energy efficiency and increase energy consumption and greenhouse gas emissions, thereby exacerbating climate change [6]. The challenges posed by climate change have prompted engineers to adopt effective measures to maintain good operation performance for various facilities within buildings, including heating, ventilation, and air conditioning (HVAC) systems, etc., while considering costs. HVAC systems usually consume more than 30% of the total energy in commercial buildings [7]. Cho suggests that in large office buildings, HVAC systems can account for 40% to 50% of the total energy consumption [8]. Chilled water networks play a vital role in transporting cooling energy within HVAC systems. However, with extended operation, these networks inevitably experience aging and wear, leading to various faults [9]. Among these, pipeline leakage is one of the most common failures throughout the lifecycle of chilled water networks. Leaks not only accelerate the corrosion of buildings but also result in the waste of water resources, as the supply water is often high-cost softened cold or hot water. Moreover, pipeline leaks may cause the system to deviate from its optimal operating point, ultimately reducing the quality of life for occupants while increasing electricity costs for HVAC systems [10]. As a large amount of electricity is generated from fossil fuels, this leads to increased greenhouse gas emissions, which is harmful to climate change mitigation [6].

Numerous studies have been conducted on pipeline leakage diagnosis both domestically and internationally. Leakage diagnosis technologies can be classified into direct and indirect methods. Direct methods involve the use of detection equipment to directly identify leaks, including acoustic methods [11], infrared thermal imaging [12], fiber optic sensing [13], and laser remote sensing [14]. In contrast, indirect methods detect leaks through changes in the pressure and flow rate by utilizing hydraulic models or algorithms for analysis. Indirect methods are further divided into two categories: model-based methods and data-driven methods. Model-based methods include pressure point methods [15], pressure gradient methods [16], negative pressure wave methods [17], state estimation methods [18], and transient modeling [19], etc. Data-driven methods that leverage machine learning techniques [20,21], such as neural networks and support vector machines, use historical sensor data to learn patterns associated with leaks. These methods can identify anomalies in pressure and flow rates, enabling leak detection without the need for precise physical models of the pipeline system.

Direct leakage detection methods are often time-consuming and costly. For instance, acoustic methods require sensors at both ends of a pipe and are effective primarily for a single long-distance pipeline, making them unsuitable for complex pipe networks [22]. Similarly, infrared thermography typically uses drones and is mainly applied to under-ground heating systems [23], which makes it impractical for indoor chilled water pipe networks. In contrast, model-based and data-driven indirect methods offer enhanced functionality, maintainability, and adaptability and enable real-time continuous detection. As a result, these methods have received increasing attention in recent research [24]. Regarding model-based approaches, Brahami et al. [25] modeled leakage as an inverse source localization problem and proposed a hybrid genetic algorithm combined with matching field processing to optimize the search for leakage sources, improving performance over traditional exhaustive search methods. Manservigi et al. [26] developed a diagnostic method based on a digital twin model to correctly diagnose common faults in district heating networks, namely leaks, heat loss, and pressure loss. Buonomano et al. [27] pointed out that white-box modeling allows for high-precision analysis and fine control. Meseguer et al. [28] established a model-driven decision support system that implements an online leakage detection and localization method based on the pressure-point technique. It has been proven to be effective for various water distribution networks. Ozgener et al. [29,30] developed an energy and energy use model to monitor district heating systems located in Afyon and Salihli by calculating energy and energy use efficiency [31]. Guan et al. [32] constructed a pressure-driven leakage model that combines leakage and pressure relationships with a nonlinear genetic algorithm by using multi-population genetic algorithms for leak localization. Kaliatka et al. [33] also detected simulated pipeline ruptures by monitoring pressure waves triggered by pressure reduction in district heating networks. For data-driven approaches, Zhou et al. [34] developed a backpropagation (BP) leakage diagnostic model for heating pipe networks, achieving an accuracy of over 89.31%. They also introduced a multi-fault diagnosis methodology based on principal component analysis (PCA) and BP neural networks for timely fault diagnosis in district heating systems [35]. Li et al. [36] proposed a leakage detection method using improved wavelet denoising and short-time Fourier transform on pressure signals, which is used in water distribution, heating networks, and gas pipeline networks. Liu et al. [10] presented a two-phase leakage fault diagnostic methodology utilizing the Adam optimization BP neural network algorithm to locate leakage in air-conditioning water pipelines. Fan et al. [37] similarly used a deep confidence network for the two-stage fault diagnosis of district heating pipe networks, which could locate leaks in pipes with an average error of 17.3 m. Kammoun et al. [38] implemented a multivariate long short-term memory (LSTM) autoencoder to create an unsupervised recurrent neural network (RNN) model for leakage detection, achieving 97% sensitivity by using leakage pressure data and 100% sensitivity with flow rate data. Wang et al. [39] conducted experiments on gas pipeline leakages. They ultimately achieved a leakage identification rate of 96.87% by using an artificial neural network with detectors and specific frequency filters for enhancing accuracy.

Despite the advances in data-driven approaches in recent years due to technological improvements, challenges still remain in practical applications. Chen et al. [40] highlighted that the interpretability of data-driven HVAC troubleshooting approaches is often inadequate, and the complexity of these systems makes it difficult to communicate their value to building owners, thus limiting their broader adoption. In addition, recent studies have applied machine learning techniques to automate fault feature extraction for leakage detection and localization in pipeline networks [10,32,34]. Although these methods achieve high detection accuracy, their black-box nature reduces interpretability and limits insights into how different leakage locations and hydraulic changes influence the system. On the other hand, although model-based methods are generally interpretable, they often rely on optimization algorithms that require iterative, time-consuming, and expensive computation. These issues present significant barriers to real-time fault diagnosis and practical deployment in engineering applications. Therefore, improving both interpretability and computational efficiency is crucial for the advancement of leakage diagnostic technologies.

To address the above limitations, this study proposes a leakage diagnosis and localization scheme tailored for closed chilled water pipe networks to overcome the limitations of acoustic and infrared thermography methods commonly used in such systems. Unlike many existing approaches that rely on iterative optimization algorithms or machine learning models requiring extensive training data, this leakage diagnosis and localization scheme utilizes pressure variation rate analysis to detect and locate leaks. This scheme offers clear interpretability as the pressure response patterns caused by leakage can be directly associated with hydraulic changes in the system. Moreover, it does not require model training or complex iterative computation, significantly reducing both diagnostic time and computational cost.

In this paper, the proposed novel leakage diagnosis and localization scheme is evaluated by using a simulation model of the hydraulic pipe network, and this model is validated by using experimental tests. The simulation hydraulic model of the pipe network under normal conditions is first established and experimentally validated. Based on this validated model, various leakage scenarios with varying leakage locations and flow rates are simulated, and pressure variation rate analysis is carried out to evaluate the effectiveness of this leakage diagnosis and localization scheme. This paper is arranged as follows. Section 2 describes the closed chilled water system for a ship-shaped building with seven zones and several user branches in each zone. Section 3 presents the leakage diagnosis scheme of the pipe network system, which is based on pressure variation rate analysis. Section 4 outlines the hydraulic models of the closed chilled water network under normal and leakage conditions. The calculation results and the leakage diagnosis of this closed pipe network system are presented in Section 5, and Section 6 concludes this paper.

2. Description of the Closed Chilled Water System

The closed chilled water system for a ship-shaped building comprises seven zones, each equipped with a pump (refrigeration unit) branch and several user branches. These user branches are distributed across the upper and lower floors, interconnected by a loop pipe, and separated by butterfly valves, allowing for an independent water supply to each zone. To simplify the analysis, this work focuses on the pipe network of Zone 1, illustrated in Figure 1. This zone includes two loops: the lower loop features six user branches, while the upper loop includes four. The pump has a flow rate of 50 m^3/h and a head of 85 mH_2O. The designed flow rates for the upper and lower loops are 20.98 m^3/h and 29.02 m^3/h, respectively, with individual branch flow rates detailed in Table 1. The pump inlet is equipped with a tank for pressure stabilization and water supply, and a balancing valve is included in the pump branch to ensure stable operation under varying conditions.

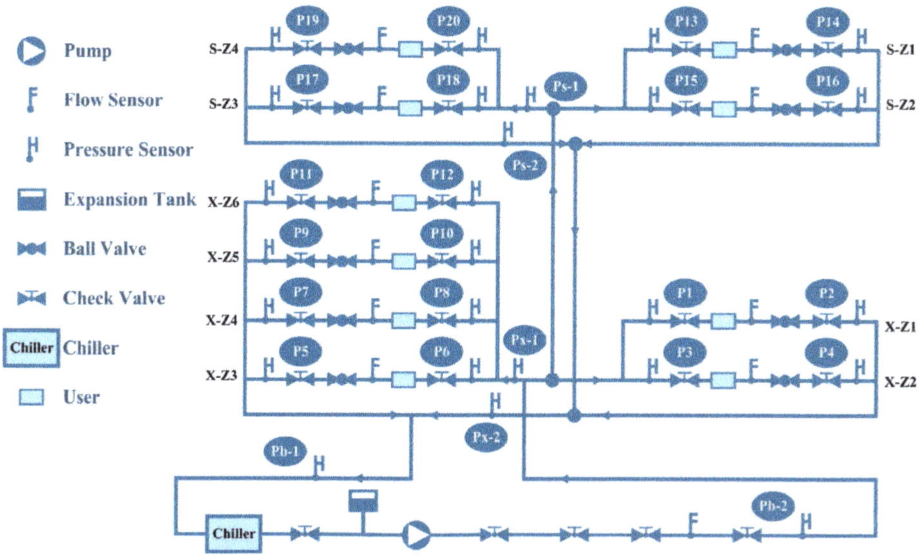

Figure 1. Schematic diagram of the Zone 1 chilled water system and measurement points.

User branches are arranged in parallel, comprising two types of end users: ordinary user branches with variable loads and electromechanical equipment users with fixed loads. Pressure sensors (Pb-1 and Pb-2) are installed at both ends of the pump branch, and pressure measurement points (Ps-1, Ps-2, Px-1, and Px-2) are positioned on the upper and lower loop pipes. The upper loop branches are designated S-Z1 to S-Z6, while the lower loop branches are labeled X-Z1 to X-Z6, with each branch equipped with a pressure point at both ends to measure pressure drops. In total, 20 pressure measurement points (P1 to P20) are established at network nodes. Pressure sensors and water flow meters are used in experimental measurements. The pressure sensor has a full-scale range of 1 MPa with an accuracy of 0.5% FS. The maximum measurement error is 0.005 MPa. The water flow meter has 1.5 times the designed flow rate of each branch, with an accuracy of 0.5% FS. For the raw measured data processing, a signal filter was used to reduce the influence of random noise, and measurement uncertainty was controlled within 0.0029 MPa for pressure measurement. The accuracy and reliability of the proposed diagnostic methodology can be ensured because the pressure values observed during the tests were generally higher than 0.15 MPa, which is significantly above the error and uncertainty range. Additionally, globe valves and ball valves are installed in each user branch to facilitate switching and balancing adjustments.

Table 1. Pipe diameters and designed flow rates of the pipeline sections of the Zone 1 chilled water system.

Zone 1 Network	Branch No.	Pipe Diameter (mm)	Designed Flow Rate (m^3/h)
Pump branch	B	80	50
Upper loop	S-Z1	20	2.60
	S-Z2	32	7.61
	S-Z3	32	5.71
	S-Z4	32	5.06
Lower loop	X-Z1	25	3.37
	X-Z2	25	4.91
	X-Z3	25	3.41
	X-Z4	20	2.92
	X-Z5	40	8.84
	X-Z6	32	5.57

3. Leakage Diagnosis Scheme of the Pipe Network System Based on Pressure Variation Rate Analysis

When the pipe network operates normally and no leakage occurs, the pressure at each node of the pipe network is constant and higher than atmospheric pressure. When a leak occurs somewhere, the negative pressure wave generated at the leakage point spreads to all parts of the system and causes changes in pressure at each node. The closer the node is to the leakage point, the greater the magnitude of the pressure variation. For a point where the measured pressure at the normal operation is p^{A}, the measured pressure is p^{B} at this point when leakage occurs at this point. The pressure variation rate at this point before and after the leakage is formulated as follows.

$$\sigma = \frac{|p^{\mathrm{A}} - p^{\mathrm{B}}|}{p^{\mathrm{A}}} \tag{1}$$

where σ is the pressure variation rate.

In order to better judge the occurrence of leakages in the pipe network system, it is necessary to set an appropriate and reasonable threshold for the pressure variation rate $\sigma_{\mathrm{Threshold}}$. When $\sigma < \sigma_{\mathrm{Threshold}}$, it is considered that there is no leakage in the pipeline. It is considered that a leakage has occurred in the pipeline when $\sigma > \sigma_{\mathrm{Threshold}}$. The specific value of σ is related to the proximity to the leakage point, as well as the size of the pipeline and the flow rate.

Referring to the "design code of district heating networks" in China [41], the accidental supply water flow rate of the closed heat network should not be less than 4% of the circulating flow rate of the heat supply system. For the closed-loop chilled water pipe network system in this paper, when the leakage amount is 4% of the total rated flow rate of the pipe network, the range of pressure variation rates at each node is investigated. Based on the maximum pressure variation rates, the threshold value of the pressure variation rate for determining the occurrence of leakages in the pipe network is determined. In practice, when the maximum value of the pressure change rates is greater than this threshold, the leakage flow rate can be considered to be more than 4% of the total rated flow rate, which is then recognized as a leakage.

The response time of leak detection is closely related to the time required for the system to reach a steady state after a leak occurs. Once the leak happens, the system stabilizes quickly, and the sensor data also reach a stable state within a short period. Based on the experiments in this study, the sensor data typically stabilize within 8 min after the leak occurs. After the data stabilize, a period of time is selected to record sensor data as steady-state data. The length of the selected time period is closely related to the sensor data's update frequency. In this study, the pressure sensors have a sampling frequency of 1 Hz, and the chosen time period (periods A and B in Figure 2) for steady-state data collection is 5 min.

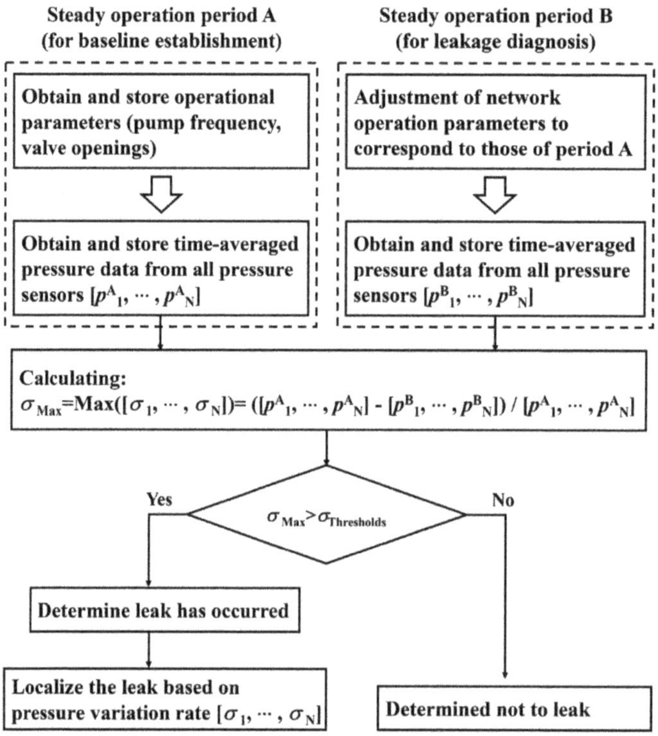

Figure 2. Schematic diagram of the leakage diagnosis scheme for the pipe network system.

Based on the above analysis, a leakage diagnosis scheme of the pipe network system is proposed, as shown in Figure 2. There are four main steps, i.e., baseline data establishment, leakage data collection, pressure variation rate calculation, and threshold comparison. They are described as follows.

Step 1: Baseline data establishment

Select a period (period A) of steady-state operation when the pipe network operates normally. Record and store the operation parameters of the pipe network, including pump frequency and valve ON/OFF states. Process all data with the time-averaged $[p^{A}{}_{1}, \cdots, p^{A}{}_{N}]$ from all pressure sensors (N is the number of pressure sensors).

Step 2: Leakage data collection

Maintain the network in steady-state operation for a subsequent period (period B). Adjust the operating parameters during this period coincident with those recorded in period A, except that a valve is open to simulate the leakage. Obtain and store the time-averaged data $[p^{B}{}_{1}, \cdots, p^{B}{}_{N}]$ from all pressure sensors.

Step 3: Pressure variation rate calculation

Calculate the pressure variation rates $[\sigma_1, \cdots, \sigma_N]$ for all sensors during period B and identify the maximum value σ_{Max} among these pressure variation rates.

Step 4: Leakage Diagnosis

Compare the maximum pressure variation rate σ_{Max} with a predefined threshold $\sigma_{Threshold}$. If $\sigma_{Max} < \sigma_{Threshold}$, it is concluded that there is no leakage in the pipe network system. Otherwise, a leakage is detected if $\sigma_{Max} > \sigma_{Threshold}$. The leakage location is identified based on the sequence of pressure variation rates $[\sigma_1, \cdots, \sigma_N]$ from all sensors.

4. Hydraulic Models of the Pipe Network and Validation

In this section, two hydraulic models of the pipe network are developed by using Modelica [42] to simulate the operation of the pipe network system. One hydraulic model (i.e., a hydraulic model for normal operation) is developed to simulate operation in normal operation conditions, and the other hydraulic model (i.e., a hydraulic model for leakage conditions) simulates operation when leakages occur in the closed chilled water pipe network. The hydraulic model for normal operations is validated against the experimental data from the in-field pipe network tests at various pump frequencies. Similarly, the hydraulic model for leakage conditions is also validated through experiments designed to assess the precision of the model under different leakage locations and flow rates.

4.1. Calculation Principle of the Pipe Network Components

In this section, the principle and mathematical models of the main components of the pipe network, i.e., the pipe, pump, valve, resistance components, and flow source, are briefly described since they are very important for developing the hydraulic models of the pipe network by using Modelica.

Pumps can be classified by their specific speed (N_s) based on their hydrodynamic properties. The specific speed [43] is a dimensionless number determined by the operation conditions of the pump, and the calculation is based on the performance index of the pump at the optimum efficiency point, with a mathematical expression as shown in Equation (2).

$$N_S = \frac{NQ^{1/2}}{H^{3/4}} \tag{2}$$

where N is the pump speed, r/min; Q is the pump flow, m^3/s; and H is the pump head, Pa.

A cylindrical rigid pipe is used as the pipe element. The fluid dynamics calculations for this pipe type are based on the Colebrook–White equation [44], as shown in Equations (3)–(6).

When $Re \leq 2000$, the fluid is in laminar flow:

$$f = f_l = \frac{64}{Re} \tag{3}$$

When 2000 < $Re \leq 4000$, the fluid is in transition:

$$f = x f_t + (1 - x) f_l \tag{4}$$

$$x = \frac{Re - 2000}{2000} \tag{5}$$

When $Re \geq 4000$, the fluid is in a turbulent state:

$$f = f_t = \frac{0.25}{\left[\log\left(\frac{k}{3.7D} + \frac{5.74}{Re^{0.9}}\right)\right]^2} \tag{6}$$

where f is the friction coefficient; f_l is the friction coefficient of laminar flow state; f_t is the friction coefficient of the turbulent flow state; Re is the Reynolds number; D is the inner diameter of the pipe, m; and k is the absolute roughness of the inner wall of the pipe, mm.

For resistance components, the resistance can be calculated by Equation (7).

$$p_1 - p_2 = \frac{K m_2 |m_2|}{2A^2 \rho} \tag{7}$$

where P_1 and P_2 are the inlet and outlet pressure values, Pa; K is the resistance coefficient; m_2 is the mass flow rate through the outlet, kg/s; A is the resistance pieces of the circulation cross-section area, m^2; and ρ is the density of the fluid, kg/m^3. By adjusting the resistance coefficient K, it is possible to change the pipe's resistance and thus distribute and adjust the flow in the pipe network.

The valve resistance calculation formula is the same as (6). However, for the resistance coefficient K, the resistance of the valve is not a preset constant value but is affected by a variety of factors such as the valve opening, fluid properties, etc. The calculation of K is shown in Equation (8).

$$K = C_{Re} K_v \tag{8}$$

where C_{Re} is the Reynolds correction factor and K_v is the valve opening factor.

In order to simulate pipe leakage, the leakage flow is simplified to a negative flow source component. The equation for the flow source model is shown in Equation (9). By setting m_{in}, the flow rate can be controlled. If it is used to model leakage, m_{in} can be set to a negative value.

$$m_{\text{flow}} = m_{\text{in}} \tag{9}$$

where m_{flow} is the flow rate of the flow source (kg/s) and m_{in} is the flow rate of the artificially set flow source (kg/s).

4.2. Hydraulic Model of the Pipe Network System for Normal Operation

The hydraulic model of the pipe network system is coded using Modelica. This hydraulic model for the closed chilled water system is developed based on the physical topology of the pipe network, i.e., the length, pipe diameter, number of valve components (including valves, tees, and elbows), and locations of pressure measurement points for each pipe section, etc. To simplify the piping system, only the pump is provided in the pump branch, and a branch resistance component is used to simulate the total resistance of the other components in the pump branch (chillers, filters, piping, etc.). In addition, a constant pressure point is provided at the pump inlet to ensure a steady water supply pressure. The model of the Zone 1 chilled water system pipe network is illustrated in Figure 3.

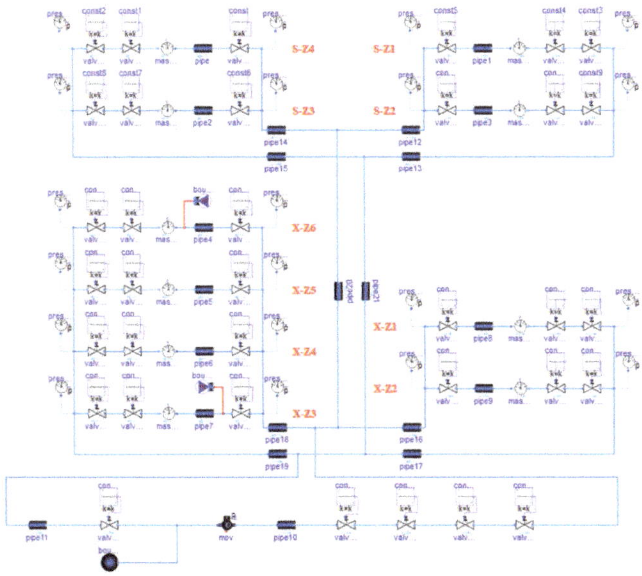

Figure 3. Hydraulic model for the normal operation of the Zone 1 cooling water pipe network system.

4.3. *Validation of the Hydraulic Model for Normal Operation*

To ensure that the hydraulic model of the pipe network for normal operation accurately represents the hydraulic characteristics of the actual system under normal operation conditions, model validation was conducted using the measured in situ operation parameters in Zone 1. Based on the measured pressure and flow rates, the branch resistance coefficient was estimated, and the resistance coefficients of each component, together with the pipe roughness and other parameters, were adjusted to align the simulated values of the coincident point with the actual measurement. The hydraulic parameters of the pipe network were further evaluated under different pump operation frequencies in the simulation. Taking Zone 1 as an example, Figure 4 presents the errors between the simulation results by using the hydraulic model for normal operation and the actual measurements. The comparison results show that the errors between the simulation and the measurements are all within 5%. It indicates that the hydraulic model for normal operation conditions can accurately represent the actual system and effectively reflect the hydraulic characteristics of the real closed chilled water network system under normal operation conditions.

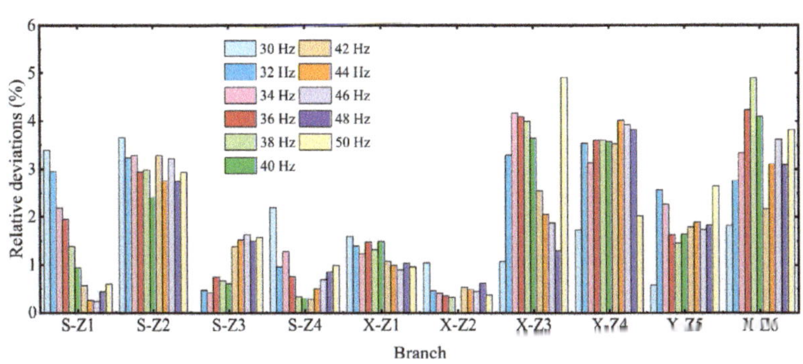

Figure 4. Relative deviations of measured and simulated pressure drops in the Zone 1 chilled water system under different pump frequencies.

4.4. Hydraulic Model of the Pipe Network System for Leakage Conditions

In practical engineering, a pipe network leakage initiates a transient dynamic process. If the damage does not worsen, the system will quickly reach a steady state. Given the numerous branches in the Zone 1 pipe network, which more accurately reflect the pressure changes in each branch, this work focuses on the leakage diagnosis of the Zone 1 pipe network under steady-state conditions, where the leakage flow rate remains constant. A flow source component (shown in Figure 5) is utilized to represent the hydraulic state during leakage. The simulation of pipe network leakages is achieved by applying a negative flow rate, with leakage points located on branches X-Z3 and X-Z6. Four leakage flow rates, denoted as f_1, f_2, f_3, and f_4, were established, accounting for 2%, 4%, 10%, and 20% of the total designed flow rate of the Zone 1 pipe network, respectively. The leakage conditions are summarized in Table 2.

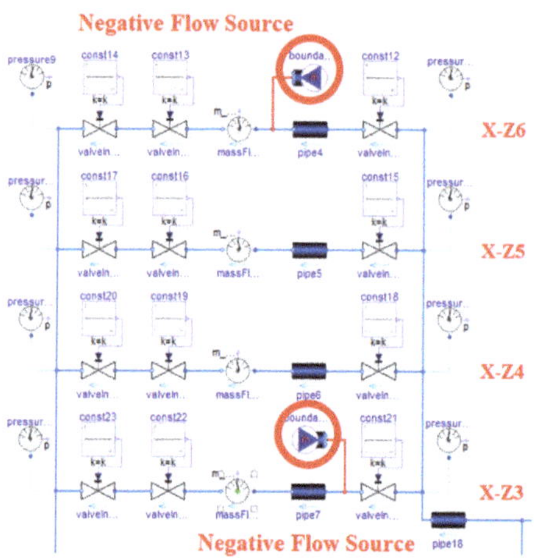

Figure 5. Hydraulic model of leaks in the network of the Zone 1 chilled water system.

Table 2. Description of leakage branches and leakage flow rates.

Branch No.	Leakage Flow f_1 for 2% (m^3/h)	Leakage Flow f_2 for 4% (m^3/h)	Leakage Flow f_3 for 10% (m^3/h)	Leakage Flow f_4 for 20% (m^3/h)
X-Z3	1	2	5	10
X-Z6	1	2	5	10

4.5. Validation of the Hydraulic Model for Leakage Conditions

To ensure that the hydraulic model of the pipe network system accurately reflects the hydraulic characteristics of the actual system under leakage conditions, model verification was conducted using measured hydraulic parameters from the pipe network in Zone 6, as illustrated in Figure 6. This zone is equipped with a pump (refrigeration unit) branch and several user branches, comprising two circuits with five user branches in the lower loop and three in the upper loop. The water pump has a flow rate of 50 m^3/h and a head of 80 mH_2O, with designed flow rates of 16.38 m^3/h for the upper loop and 33.62 m^3/h for the lower loop, as detailed in Table 3.

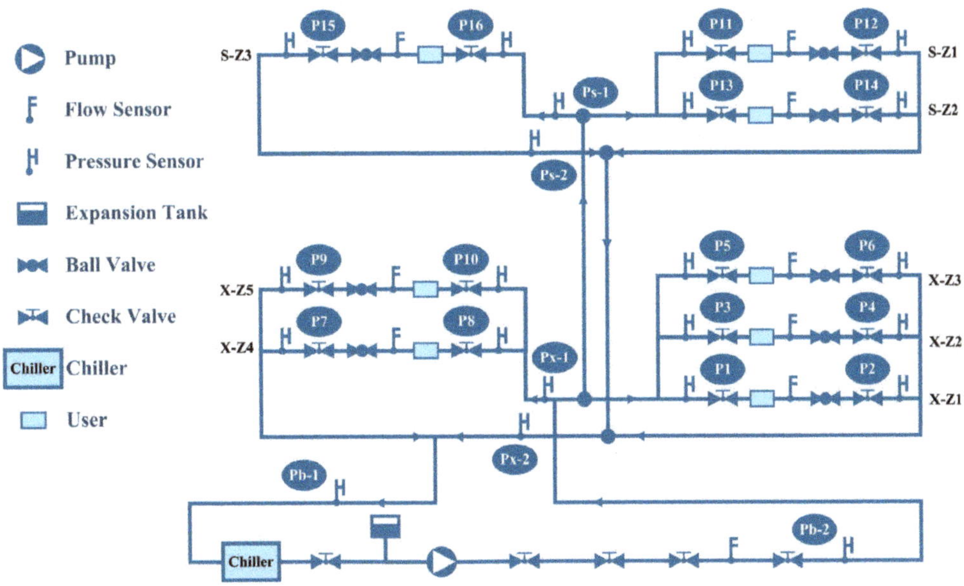

Figure 6. Schematic diagram of the Zone 6 chilled water system and measurement points.

Table 3. Pipe diameters and designed flow rates of pipeline sections in the Zone 6 chilled water system.

Zone 6 Network	Branch No.	Pipe Diameter (mm)	Designed Flow Rate (m^3/h)
Pump branch	A6-B	32	50
Upper loop	A6-S-Z1	25	5.89
	A6-S-Z2	32	3.17
	A6-S-Z3	25	7.32
Lower loop	A6-X-Z1	32	3.74
	A6-X-Z2	25	5.96
	A6-X-Z3	32	4.05
	A6-X-Z4	50	6.22
	A6-X-Z5	32	13.65

The inlet of the water pump is connected to a supply pipe leading to a tank, which stabilizes the pressure and provides water during the leakage experiment. This supply pipe includes a supply pump and a flow sensor. Additionally, pressure sensors are installed at both ends of the pump branch (i.e., pressure point B-P1 on the water supply side and B-P2 on the return side). The upper loop branches are labeled S-Z1 to S-Z3, while the lower loop branches are labeled X-Z1 to X-Z5. Each branch is equipped with a pressure point and a flow sensor to measure the pressure and flow rates at both the supply and return ends.

The leakage conditions for the site experiment are outlined in Table 4, detailing leakage locations and flow rates. The model validation process is as follows. First, a simulation model of the Zone 6 pipe network was established (see Figure 7), and the resistance coefficients, pipe roughness, and other parameters were adjusted based on measured values during normal operating conditions, as described in Section 2. Hydraulic calculations were then performed under various leakage scenarios. The leakage simulation model is depicted in Figure 8.

Table 4. Tests on different leakage locations and leakage flow rates.

Leakage Test No.	Leakage Test 1	Leakage Test 2	Leakage Test 3	Leakage Test 4
Leakage branch	S-Z2	X-Z2	S-Z2	X-Z2
Leakage position	Supply side	Supply side	Return side	Return side
Leakage flow (m^3/h)	0.2	0.58	0.2	0.3

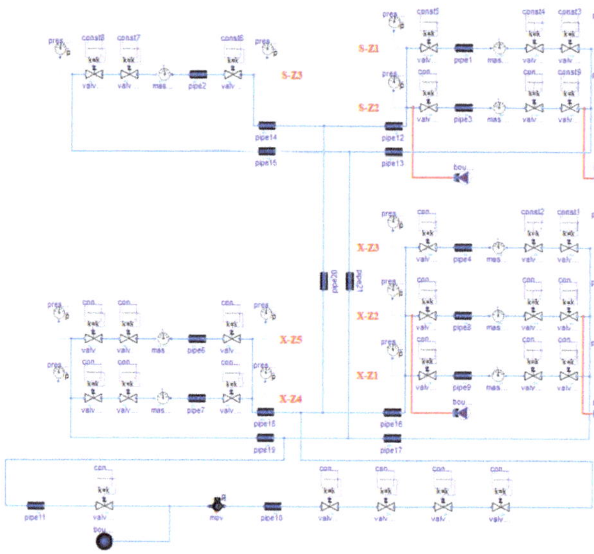

Figure 7. Simulation model of the pipe network in the Zone 6 chilled water system.

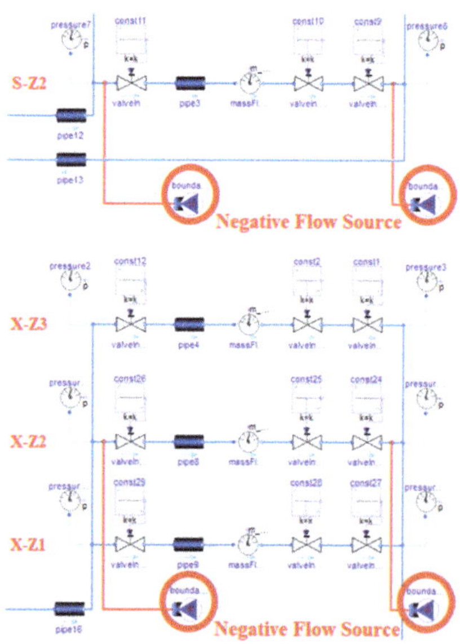

Figure 8. Hydraulic model of leaks in the network of the Zone 6 chilled water system.

As shown in Figure 9, the errors between the simulation results and actual measurements are all within 4%, indicating that the model closely aligns with the actual system and accurately reflects the hydraulic state under leakage conditions.

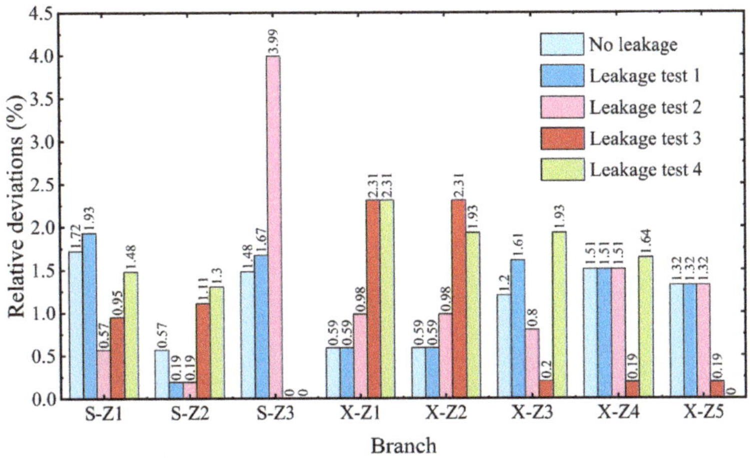

Figure 9. Relative deviations of measured and simulated pressure drops in the Zone 6 chilled water system.

5. Calculations Results and Leakage Diagnosis of the Closed Pipe Network System

Zone 1 of the closed chilled water pipe network system was used to simulate the leakage and validate the leakage diagnosis scheme since this zone has a higher density and a greater number of branches. This zone allows for a more comprehensive analysis of pressure variation rates at each measurement point when a leakage occurs. The selected leakage locations are depicted in Figure 5. A leakage point is set up on the water supply side (supply side of branch X-Z3) and on the return side (return side of branch X-Z6) of the different branches, respectively. This arrangement enables us to analyze the distinct patterns of leakages occurring on the water supply side versus the water return side. It is worth noting that only one leakage point occurrence at a time is considered since two or more leakages rarely occur simultaneously in typical scenarios. Therefore, the diagnosis and analysis are confined to examining one point leakage at a time to ensure the clarity of the results.

5.1. Leakage Diagnosis at the Supply End of the Branch

The leakage is imposed on the water supply side of the X-Z3 branch at the entrance of the terminal user, as shown in Figure 5. Following the approach outlined in Section 4.4, a negative flow source is used to simulate the leakage.

Figure 10 illustrates the variation rate of the pressure monitoring point when the leakage point is positioned near the water supply end of the X-Z3 branch and different leakage levels are imposed. Four leakage flow levels are considered, as shown in Table 2, representing 2%, 4%, 10%, and 20% of the total rated flow of the pipe network. It shows that there is a corresponding pressure decrease at both the supply end (pressure points P1, P3, etc.) and the return end (pressure points P2, P4, etc.) of all branches in the system. For the leakage flow rate f_1 of 1 m^3/h, the pressure variation rate at the monitoring point P6 (located at the water supply end of X-Z3) is the highest at 1.18%, followed closely by adjacent branches (P8, P10, and P12) with rates of approximately 1.16%. When the flow rate increases to f_2 (2 m^3/h), P6 again shows the highest variation rate at 2.25%, with adjacent branches displaying rates around 2.23%. At a flow rate f_3 of 5 m^3/h, P6 records a variation rate of 5.86%, while the adjacent branches show rates of about 5.73%. For a flow rate f_4 of

10 m^3/h, P6 reaches a peak variation rate of 12.27%, followed by adjacent branches with rates near 11.86%.

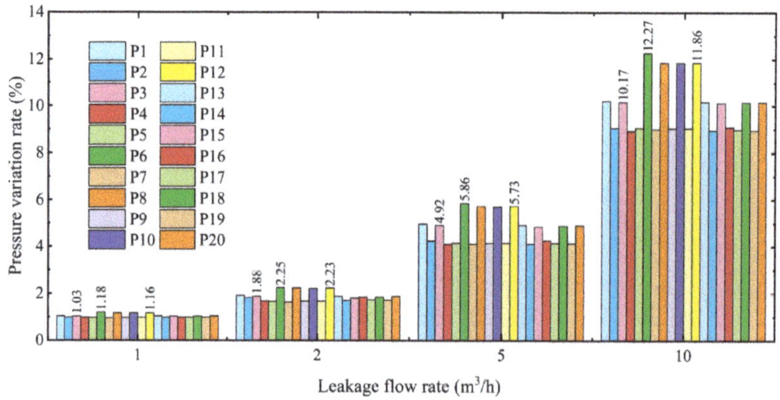

Figure 10. Pressure variation rate of each monitoring point with leakage occurrence in the branch of X-Z3.

Figure 10 also shows that of the four leakage levels for different flow rates, pressure point P6 at the water supply end of the leakage branch always has the highest pressure variation rate. The pressure variation rates at monitoring points other than P6 can be categorized into three levels according to their magnitude. The first level includes the pressure monitoring points at the water supply end of the loop, where the leakage branch is located, except for the leakage branch (P8, P10, and P12); the second level includes the pressure monitoring points at the water supply end of each branch in the loops, except for the loop where the leakage branch is located (P1, P3, P13, P15, P18, and P20); and the third level includes the pressure monitoring points at the return end of all branches (P2, P4, P5, P7, P9, P11, P14, P16, P17, and P19).

The above pattern can be used to locate leakage. We sorted the pressure variation rates of all the pressure points and found that the pressure variation rates of the monitoring points located in water supply monitoring are generally greater than the pressure variation rates of the monitoring points on the return side, which can be determined to be caused by a leakage in the water supply side. The branch with the highest possibility of leakage is the branch with the largest pressure variation rate, P6. This is followed by P8, P10, and P12, which are located at the water supply end of the loop in which the leakage branch is located, except for the leakage branch. In fact, the location of the leakage is indeed the water supply end of the branch where P6 is located.

5.2. Leakage Diagnosis at the Return End of the Branch

The leakage is imposed on the return side of the X-Z6 branch, at the outlet of the terminal user, as shown in Figure 5. Following the methodology outlined in Section 4.4, a negative flow source was employed to simulate the leakage.

Figure 11 illustrates the variation rate of the pressure monitoring point when the leakage point is positioned near the return end of the branch of X-Z6 and different leakage levels are imposed. Four leakage flow levels are considered, as shown in Table 2, representing 2%, 4%, 10%, and 20% of the total rated flow of the pipe network. It shows there is a corresponding pressure decrease at both the supply end (pressure points P1, P3, etc.) and the return end (pressure points P2, P4, etc.) of all branches in the system. For the leakage flow rate f_1 of 1 m^3/h, the pressure variation rate at the monitoring point P11, situated at the return end of branch X-Z6, is the highest at 1.89%, followed closely by the pressure variation rates at the return ends of adjacent branches (P9, P7, and P5), which are approximately 1.80%. When the leakage flow rate f_2 is 2 m^3/h, P11 again exhibits the largest variation rate at 4.25%, with adjacent branches showing rates around 4.10%. At a flow rate f_3 of 5 m^3/h, the pressure variation rate at P11 rises to 10.41%, while adjacent branches have rates of about 10.07%. For a flow rate f_4 of 10 m^3/h, P11 records the highest variation rate at 20.27%, with adjacent branches displaying rates near 19.65%.

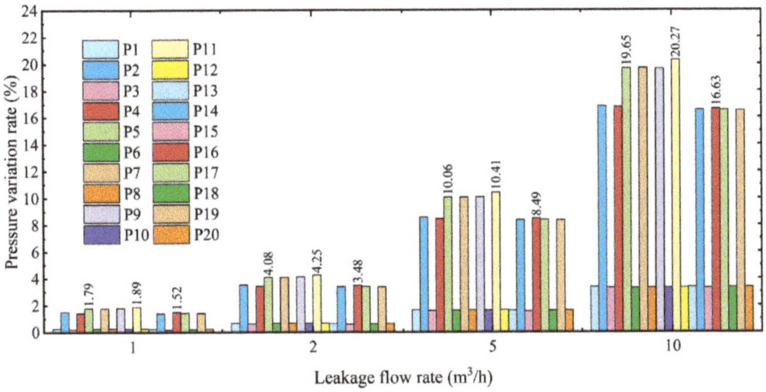

Figure 11. Pressure variation rate of each monitoring point with leakage occurrence in the branch of X-Z6.

Figure 11 also shows that of the four leakage levels for different flow rates. Pressure point P11 at the return end of the leakage branch always has the highest pressure variation rate. The pressure variation rates at the other monitoring points can be categorized into three levels according to their magnitude. The first level includes the pressure monitoring points at the return end of the loop where the leakage branch is located, except for the leakage branch (P5, P7, and P9); the second level includes the pressure monitoring points at the return end of each branch in the loops, except for the loop where the leakage branch is located (P2, P4, P14, P16, P17, and P19); and the third level includes the pressure monitoring points at the supply end of all branches (P1, P3, P6, P8, P10, P12, P13, P15, P18, and P20).

The above pattern can be used to locate leakage. We sorted the pressure variation rates of all the pressure points and found that the pressure variation rates of the monitoring points located in the return monitoring are generally greater than the pressure variation rates of the monitoring points on the supply side, which can be determined to be from a leakage in the return side. The branch with the highest possibility of leakage is the branch with the largest pressure variation rate, P11. This is followed by P5, P7, and P9, which are located at the return end of the loop in which the leakage branch is located, except for the leakage branch. In fact, the location of the leakage is indeed the return end of the branch where P11 is located.

5.3. Discussion of the Threshold Selection of the Pressure Variation Rate

According to Section 3, with reference to the "design code of district heating networks" in China [41], it is generally accepted in engineering that leakage is identified when the leakage volume exceeds 4% of the total designed flow of the pipe network. For instance, in the examined pipe network, a leakage flow rate of 2 m^3/h (representing 4% of the rated designed flow) results in a pressure variation rate of approximately 3% at the monitoring point. When the leakage flow rate exceeds 2 m^3/h (e.g., rates f_3 and f_4, with leakage flow rates of 5 m^3/s and 10 m^3/s, respectively), the pressure variation rate at the measurement point increases significantly beyond 3%. Thus, a pressure variation rate of 3% can be used as the threshold for detecting the occurrence of leakage within the closed chilled water pipe network system. When the maximum variation rate of the pressure measurement points in the pipe network is found to be more than 3%, it can be determined that there is a leakage in the pipe network. When leakage is detected in the pipe network, the pressure variation rates of all pressure monitoring points are ranked in order of magnitude. If the pressure variation rate of the pressure monitoring points located on the water supply side (return side) is generally greater than that on the return side (water supply side), then it can be determined that leakage has occurred on the water supply side (return side) of the pipe network. The branch where the pressure monitoring point with the largest pressure variation rate is located is most likely to be the leakage branch, from which the leakage location can be realized.

5.4. Discussion of the Practical Feasibility and Scalability of the Diagnosis Scheme

To ensure the practical feasibility of this novel leakage diagnosis and location scheme, the relationship between sensor measurement errors and the overall diagnostic accuracy is investigated and evaluated. Many studies have indicated that sensor accuracy exerts only a minor influence on leakage diagnosis [22]. In addition, data preprocessing and noise filtering techniques can effectively reduce the impact of sensor noise on system measurement data, thereby enhancing the reliability of the measurements [45]. In this study, the novel leakage diagnosis and location scheme hinges solely on quantitative comparisons of the relative magnitude of sensor data, rather than on absolute pressure measurement. As a result, minor measurement errors do not obviously influence the data ranking of the pressure variation rate or the leak localization process. Specifically, when the detected pressure change rate exceeds 3%, a leakage is deemed to have occurred. The threshold of 3% is substantially higher than the sensor's accuracy of 0.5% FS, which typically corresponds to leakage rates above 2 m^3/h. Notably, the small sensor error does not materially affect the final determination of the leak.

The response time of the diagnosis scheme is also a key factor in its practical implementation. The response time mainly depends on the data transmission speed of the monitoring system and the computational speed of the diagnosis model. In practice, the data acquisition and transmission system operates with minimal delay (usually within a few milliseconds), ensuring real-time data availability. Meanwhile, the proposed diagnosis scheme utilizes a non-iterative algorithm with low computational complexity, which avoids the time-consuming iterative computation common to other schemes. As a result, the proposed scheme allows for rapid leakage diagnosis.

In terms of scalability, the proposed scheme relies on the relative magnitude of pressure variation rates between sensors rather than complex global data fitting or absolute precision. Additionally, it does not require iterative or training computations. As a result, the scheme can be transferred to other pipe networks without the need for re-modeling or re-training, offering a certain degree of scalability. However, to ensure effective deployment, a reasonable sensor layout is still required to capture the key pressure variation relationships.

6. Conclusions

Efficient leakage diagnosis is crucial for the timely management of faults in chilled water pipe systems, minimizing damage and ensuring the safe and stable operation of the network. This paper proposes a leak diagnosis and localization scheme based on the analysis of pressure variation rates. Hydraulic models of the pipeline network under both normal and leakage conditions are established and validated by experimental tests. By simulating the operation of the pipe network under various leakage points and flow rates, the relationship between the pressure variation rate at each node (branch end) and different leakage conditions is identified. This analysis allows for the determination of the leakage point (branch) based on the observed pressure variation rate patterns. Moreover, to accurately assess leakage, a reasonable threshold for the pressure variation rate must be established. Using a 4% leakage flow rate as a benchmark, a range of pressure variation rates is derived for each node, and a threshold is set to recognize the occurrence of leakage. The key conclusions are as follows.

(1). Taking the case of a 20% leakage flow rate of the total rated flow rate as an example, when a leak occurs at the water supply end (return end) of a branch, the pressure variation rate at the measurement point at the water supply end (return end) of the affected branch is the highest, measuring 12.27% (20.27%). This indicates the branch location of the leak. The pressure variation rates of the remaining measurement points can be categorized into three levels. The first level includes the pressure monitoring points at the water supply end (return end) of the loop where the leakage branch is located, except for the leakage branch, with rates around 11.86% (19.65%); the second level includes the pressure monitoring points at the water supply end (return end) of each branch in the loops, except for the loop where the leakage branch is located, with rates around 10.22% (16.83%); and the third level includes the pressure monitoring points at the return end (water supply end) of all branches, with rates around 9.06% (3.36%).

(2). The pressure variation rate at each node of the pipe network increases with the leakage flow rate. In this study, as the proportion of leakage flow increases from 2% to 20% of the total design flow, the maximum pressure change rate at the measurement points rises from 0.411% to 20.27%. Referring to the guidelines from the "design code of district heating networks" in China, a pressure variation rate of 3% can be used as the threshold for detecting the occurrence of leakage within the closed chilled water pipe network system.

(3). When leakage is detected in the pipe network, the pressure variation rates of all pressure monitoring points are ranked in order of magnitude. If the pressure variation rate of the pressure monitoring points located on the water supply side (return side) is generally greater than that on the return side (water supply side), then it can be determined that leakage has occurred on the water supply side (return side) of the pipe network. The branch where the pressure monitoring point with the largest pressure variation rate is located is most likely to be the leakage branch, from which the leakage location can be realized.

Author Contributions: Conceptualization, X.Z.; methodology, X.Z.; software, L.L. and S.P.; validation, S.P.; formal analysis, L.L.; investigation, X.Z. and L.L.; resources, F.L. and J.X; data curation, X.Z.; writing—original draft preparation, X.Z. and S.P.; writing—review and editing, L.L.; visualization, L.L.; supervision, J.X.; project administration, F.L. and J.X.; funding acquisition, F.L. All authors have read and agreed to the published version of the manuscript.

Funding: This research received no external funding.

Data Availability Statement: The original contributions presented in the study are included in the article, further inquiries can be directed to the corresponding author.

Conflicts of Interest: The authors declare no conflict of interest.

References

1. Almobarek, M.; Mendibil, K.; Alrashdan, A. Predictive Maintenance 4.0 for Chilled Water System at Commercial Buildings: A Systematic Literature Review. *Buildings* **2022**, *12*, 1229. [CrossRef]
2. Kumar, P.; Martani, C.; Morawska, L.; Norford, L.; Choudhary, R.; Bell, M.; Leach, M. Indoor air quality and energy management through real-time sensing in commercial buildings. *Energy Build.* **2016**, *111*, 145–153. [CrossRef]
3. Monge-Barrio, A.; Gutiérrez, A.S.O. *Passive Energy Strategies for Mediterranean Residential Buildings: Facing the Challenges of Climate Change and Vulnerable Populations*; Springer: Berlin/Heidelberg, Germany, 2018.

4. Yau, Y.; Hasbi, S. A review of climate change impacts on commercial buildings and their technical services in the tropics. *Renew. Sustain. Energy Rev.* **2013**, *18*, 430–441. [CrossRef]
5. Kutty, N.A.; Barakat, D.; Darsaleh, A.O.; Kim, Y.K. A Systematic Review of Climate Change Implications on Building Energy Consumption: Impacts and Adaptation Measures in Hot Urban Desert Climates. *Buildings* **2024**, *14*, 13. [CrossRef]
6. Elmouatamid, A.; Fricke, B.; Sun, J.; Pong, P.W.T. Air Conditioning Systems Fault Detection and Diagnosis-Based Sensing and Data-Driven Approaches. *Energies* **2023**, *16*, 4721. [CrossRef]
7. Li, X.; Li, Y.; Seem, J.E.; Li, P. Dynamic modeling and self-optimizing operation of chilled water systems using extremum seeking control. *Energy Build.* **2013**, *58*, 172–182. [CrossRef]
8. Cho, J.; Kim, Y.; Koo, J.; Park, W. Energy-cost analysis of HVAC system for office buildings: Development of a multiple prediction methodology for HVAC system cost estimation. *Energy Build.* **2018**, *173*, 562–576. [CrossRef]
9. Li, Y.; O'Neill, Z. A critical review of fault modeling of HVAC systems in buildings. *Build. Simul.* **2018**, *11*, 953–975. [CrossRef]
10. Liu, R.; Zhang, Y.; Li, Z. Leakage Diagnosis of Air Conditioning Water System Networks Based on an Improved BP Neural Network Algorithm. *Buildings* **2022**, *12*, 610. [CrossRef]
11. Zaman, D.; Tiwari, M.K.; Gupta, A.K.; Sen, D. A review of leakage detection strategies for pressurised pipeline in steady-state. *Eng. Fail. Anal.* **2020**, *109*, 104264. [CrossRef]
12. Murvay, P.-S.; Silea, I. A survey on gas leak detection and localization techniques. *J. Loss Prev. Process Ind.* **2012**, *25*, 966–973. [CrossRef]
13. Rui, D.; Li, J.Y.; Na, L.; Zhu, X.F.; Yang, X.J.; Chang, T. Application of distributed optical fiber temperature sensing system in heat network leakage monitoring. In Proceedings of the Annual Conference of Chinese-Society-of-Optical-Engineering (CSOE)—Optical Fiber Sensors and Communication (AOPC), Beijing, China, 5–8 August 2019.
14. Huang, W.; Li, X.Y.; Gao, X.M.; Fan, H.; Li, W.Z.; Huang, T.; Pei, S.X.; Shao, J.; Deng, L.H.; Zhang, W.J. Portable remote sensor of methane leakage using near-IR diode laser. In Proceedings of the Conference on Optical Technologies for Atmospheric, Ocean, and Environmental Studies, Beijing, China, 18–22 October 2004.
15. Behari, N.; Sheriff, M.Z.; Rahman, M.A.; Nounou, M.; Hassan, I.; Nounou, H. Chronic leak detection for single and multiphase flow: A critical review on onshore and offshore subsea and arctic conditions. *J. Nat. Gas Sci. Eng.* **2020**, *81*, 1451–1460. [CrossRef]
16. Lu, H.; Iseley, T.; Behbahani, S.; Fu, L. Leakage detection techniques for oil and gas pipelines: State-of-the-art. *Tunn. Undergr. Space Technol.* **2020**, *98*, 103249. [CrossRef]
17. Liu, B.; Jiang, Z.; Nie, W.; Wen, H. Application of VMD in Pipeline Leak Detection Based on Negative Pressure Wave. *J. Sens.* **2021**, *2021*, 8699362. [CrossRef]
18. Torres, L.; Jiménez-Cabas, J.; González, O.; Molina, L.; López-Estrada, F.-R. Kalman Filters for Leak Diagnosis in Pipelines: Brief History and Future Research. *J. Mar. Sci. Eng.* **2020**, *8*, 173. [CrossRef]
19. Hachem, F.E.; Schleiss, A.J. A review of wave celerity in frictionless and axisymmetrical steel-lined pressure tunnels. *J. Fluids Struct.* **2011**, *27*, 311–328. [CrossRef]
20. Abdulshaheed, A.; Mustapha, F.; Ghavamian, A. A pressure-based method for monitoring leaks in a pipe distribution system: A Review. *Renew. Sust. Energ. Rev.* **2017**, *69*, 902–911. [CrossRef]
21. Shen, Y.; Chen, J.; Fu, Q.; Wu, H.; Wang, Y.; Lu, Y. Detection of District Heating Pipe Network Leakage Fault Using UCB Arm Selection Method. *Buildings* **2021**, *11*, 275. [CrossRef]
22. Yang, G.; Xing, D.; Wang, H. Leak localization in District Heating Networks integrating physical model-based and data driven-based methods: Impact of dataset construction on model performance. *Energy* **2024**, *308*, 132839. [CrossRef]
23. Zhou, S.; O'Neill, Z.; O'Neill, C. A review of leakage detection methods for district heating networks. *Appl. Therm. Eng.* **2018**, *137*, 567–574. [CrossRef]
24. Xue, P.; Jiang, Y.; Zhou, Z.; Chen, X.; Fang, X.; Liu, J. Machine learning-based leakage fault detection for district heating networks. *Energ. Buildings* **2020**, *223*, 110161. [CrossRef]
25. Brahami, M.A.; Abdi, S.M.; Hamdi Cherif, S.; Bendahmane, A. Optimization of a Pipelines Leak Detection Method Based on Inverse Transient Analysis Using a Genetic Algorithm. *Arab. J. Sci. Eng.* **2022**, *48*, 1451–1460. [CrossRef]
26. Manservigi, L.; Bahlawan, H.; Losi, E.; Morini, M.; Spina, P.R.; Venturini, M. A diagnostic approach for fault detection and identification in district heating networks. *Energy* **2022**, *251*, 123988. [CrossRef]
27. Buonomano, A.; Forzano, C.; Mongibello, L.; Palombo, A.; Russo, G. Optimising low-temperature district heating networks: A simulation-based approach with experimental verification. *Energy* **2024**, *304*, 131954. [CrossRef]
28. Meseguer, J.; Mirats-Tur, J.M.; Cembrano, G.; Puig, V.; Quevedo, J.; Pérez, R.; Sanz, G.; Ibarra, D. A decision support system for on-line leakage localization. *Environ. Modell. Softw.* **2014**, *60*, 331–345. [CrossRef]
29. Ozgener, L.; Hepbasli, A.; Dincer, I. Energy and exergy analysis of geothermal district heating systems: An application. *Build. Environ.* **2005**, *40*, 1309–1322. [CrossRef]
30. Ozgener, L.; Hepbasli, A.; Dincer, I. Energy and exergy assessment of Salihli geothermal district heating system. In Proceedings of the World Geothermal Congress 2005, Antalya, Turkey, 24–29 April 2005.

31. Ozgener, L.; Ozgener, O. Monitoring of energy exergy efficiencies and exergoeconomic parameters of geothermal district heating systems (GDHSs). *Appl. Energy* **2009**, *86*, 1704–1711. [CrossRef]
32. Guan, Y.; Lv, M.; Dong, S. Pressure-driven Background Leakage Models and their Application for Leak Localization Using a Multi-population Genetic Algorithm. *Water Resour. Manag.* **2022**, *37*, 359–373. [CrossRef]
33. Kaliatka, A.; Valincius, M. Modeling of pipe break accident in a district heating system using RELAP5 computer code. *Energy* **2012**, *44*, 813–819. [CrossRef]
34. Zhou, S.; Liu, C.; Zhao, Y.; Zhang, G.; Zhang, Y. Leakage diagnosis of heating pipe-network based on BP neural network. *Sustain. Energy Grids Netw.* **2022**, *32*, 100869. [CrossRef]
35. Zhou, S.; Liu, X.; Tian, Y.; Zhang, C.; Li, F.; Jiang, G. Multi-fault diagnosis of district heating system based on PCA_BP neural network. *Process Saf. Environ. Protect.* **2024**, *186*, 301–317. [CrossRef]
36. Li, H.; Li, H.; Pei, H.; Li, Z. Leakage detection of HVAC pipeline network based on pressure signal diagnosis. *Build. Simul.* **2019**, *12*, 617–628. [CrossRef]
37. Fan, Q.; Guo, Y.; Wu, S.; Liu, X. Two-Level Diagnosis of Heating Pipe Network Leakage Based on Deep Belief Network. *IEEE Access* **2019**, *7*, 182983–182992. [CrossRef]
38. Kammoun, M.; Kammoun, A.; Abid, M. LSTM-AE-WLDL: Unsupervised LSTM Auto-Encoders for Leak Detection and Location in Water Distribution Networks. *Water Resour. Manag.* **2022**, *37*, 731–746. [CrossRef]
39. Wang, W.; Mao, X.; Liang, H.; Yang, D.; Zhang, J.; Liu, S. Experimental research on in-pipe leaks detection of acoustic signature in gas pipelines based on the artificial neural network. *Measurement* **2021**, *183*, 109875. [CrossRef]
40. Chen, Z.; O'Neill, Z.; Wen, J.; Pradhan, O.; Yang, T.; Lu, X.; Lin, G.; Miyata, S.; Lee, S.; Shen, C.; et al. A review of data-driven fault detection and diagnostics for building HVAC systems. *Appl. Energy* **2023**, *339*, 121030. [CrossRef]
41. *CJJ 34-2002*; Design Code of District Heating Network. Ministry of Construction of the People's Republic of China: Beijing, China, 2002. (In Chinese)
42. Otter, M.; Reiner, M.; Tobolář, J.; Gall, L.; Schäfer, M. Towards Modelica Models with Credibility Information. *Electronics* **2022**, *11*, 2728. [CrossRef]
43. Jafarzadeh, B.; Hajari, A.; Alishahi, M.M.; Akbari, M.H. The flow simulation of a low-specific-speed high-speed centrifugal pump. *Appl. Math. Model.* **2011**, *35*, 242–249. [CrossRef]
44. Colebrook, C.F.; White, C.M. Experiments with fluid friction in roughened pipes. *Proc. R. Soc. Lond. Ser. A-Math. Phys. Eng. Sci.* **1937**, *161*, 367–381.
45. Bahlawan, H.; Ferraro, N.; Gambarotta, A.; Losi, E.; Manservigi, L.; Morini, M.; Saletti, C.; Spina, P.R.; Venturini, M. Detection and identification of faults in a District Heating Network. *Energy Conv. Manag.* **2022**, *266*, 115837. [CrossRef]

buildings

MDPI

Article

Archetype Identification and Energy Consumption Prediction for Old Residential Buildings Based on Multi-Source Datasets

Chengliang Fan [1], Rude Liu [1] and Yundan Liao [2,*]

[1] School of Architecture and Urban Planning, Guangzhou University, Guangzhou 510006, China; chengliang.fan@gzhu.edu.cn (C.F.); urude3102@e.gzhu.edu.cn (R.L.)
[2] School of Civil Engineering and Transportation, Guangzhou University, Guangzhou 510006, China
* Correspondence: ydliao@gzhu.edu.cn

Abstract

Assessing energy consumption in existing old residential buildings is key for urban energy conservation and decarbonization. Previous studies on old residential building energy assessment face challenges due to data limitations and inadequate prediction methods. This study develops a novel approach integrating building energy simulation and machine learning to predict large-scale old residential building energy use using multi-source datasets. Using Guangzhou as a case study, open-source building data was collected to identify 31,209 old residential buildings based on age thresholds and areas of interest (AOIs). Key building form parameters (i.e., long side, short side, number of floors) were then classified to identify residential archetypes. Building energy consumption data for each prototype was generated using EnergyPlus (V23.2.0) simulations. Furthermore, XGBoost and Random Forest machine learning algorithms were used to predict city-scale old residential building energy consumption. Results indicated that five representative prototypes exhibited cooling energy use ranging from 17.32 to 21.05 kWh/m^2, while annual electricity consumption ranged from 60.10 to 66.53 kWh/m^2. The XGBoost model demonstrated strong predictive performance (R^2 = 0.667). SHAP (Shapley Additive Explanations) analysis identified the Building Shape Coefficient (BSC) as the most significant positive predictor of energy consumption (SHAP value = 0.79). This framework enables city-level energy assessment for old residential buildings, providing critical support for retrofitting strategies in sustainable urban renewal planning.

Keywords: old residential buildings; archetype; building energy; prediction model; multi-source data

1. Introduction

In 2021, urban populations accounted for 56% of the global total, and this proportion is projected to rise to 68% by 2050 [1,2]. Urban buildings account for approximately 30% of total energy consumption and contribute to 26% of greenhouse gas emissions [3]. By the end of 2024, China's permanent urbanization rate reached 67% [4], with urban residential floor area totaling 33.1 billion m^2 [5]. A substantial portion of residential buildings constructed during the initial stages of urbanization (1980–2000) consume substantially more energy than modern structures. Implementing energy efficiency retrofits in these old residential buildings is essential for achieving decarbonization targets [6,7].

To achieve urban building energy savings, conducting energy consumption assessments of old residential buildings is essential [8]. Existing methodologies for urban building energy modeling are primarily categorized as top-down or bottom-up approaches [9]. Top-down approaches estimate energy use at the sector level [10], utilizing statistical models, regression analyses, building stock data, technology adoption models, and economic models. However, they typically lack precise temporal or spatial resolution results of individual building energy consumption, and it is often difficult to obtain dynamic data [11,12]. In contrast, the bottom-up approach is usually used to calculate energy consumption for individual buildings. Bottom-up methodologies can be classified into three types based on application method: statistical models, physical models, and hybrid models [13]. Statistical models establish correlations between building characteristics, energy usage, and socio-economic indicators. Physical models, conversely, compute energy consumption by simulating a building's physical attributes (e.g., insulation, orientation) and technical specifications (e.g., HVAC systems), though this process demands substantial computational resources. The hybrid approach integrates the advantages of both methods, offering heightened flexibility across diverse building energy application scenarios [9].

Obtaining city-scale building data presents significant challenges, hindering comprehensive energy consumption evaluation at this scale [14]. The inherent complexity of modeling energy use for individual buildings further compounds this difficulty. Prototype energy modeling is regarded as an effective method for large-scale building energy assessment [15–17]. This approach can be implemented using building codes-based, data-driven, and hybrid methods [18]. EnergyPlus [19] and DeST [20] are the predominant simulation tools for prototype buildings. Numerous studies establish prototype models to assess specific building energy. For example, Carnieletto et al. [21] developed 46 prototypes (16 single-family, 16 multi-family, 14 office buildings) for energy consumption. Liu and Gou [22] established six residential building prototypes and an energy consumption model, identifying climate, population, and income as significant Energy Use Intensity (EUI) influencers. Deng et al. [23] constructed 66 prototype models for Changsha city from geographic information system (GIS) data (68,966 buildings), covering residential, commercial, and office types. They subsequently employed clustering and Random Forest (RF) models to identify 22 representative prototypes and analyze their energy consumption [24]. Song et al. [25] built various prototypes based on multi-source open data (such as maps and satellite images) and used simulated energy intensity with EnergyPlus. An et al. [26] developed 151 building prototypes for China across five climate zones and construction periods aligned with national standards. Peng et al. [27] constructed seven typical residential prototypes through spatial data and envelope structure of 120 residential areas (involving 1300 buildings, 900 units, 350 types), referring the latest energy-saving specifications to support energy prediction models. Xia et al. [28] developed eight prototype residential units in Guangzhou using population, urban residential data, and basic unit information. In addition, Yu [29] created a community-scale residential energy performance model based on a prototype, applied to 1963 Shanghai communities, identifying plot ratio and building coverage ratio as key influencing factors. Similarly, Alasmar et al. [30] built 21 prototype models to assess the energy consumption of Jordan's national housing stock based on available building datasets. Li et al. [31] developed a residential building shape prototype via satellite images and clustering for urban-scale building stock energy calculation. Previous studies indicate that prototype models are widely used to assess old residential building energy. However, their ability to characterize city-scale energy performance remains limited under variable data conditions.

Machine learning has demonstrated significant efficacy in city-scale building energy assessment [32,33], utilizing algorithms such as extreme gradient boosting (XGBoost), Random Forest (RF), Artificial Neural Network (ANN), Gradient Boosting Decision Tree (GBDT), and Support Vector Regression (SVR). Notably, the XGBoost model often outperforms conventional regression methods in predicting residential building loads [34] and energy consumption [35]. Machine learning models can capture nonlinear relationships between urban morphological parameters and energy performance [36–39]. RF exhibits strengths in energy consumption prediction [40] especially when leveraging multi-source data (e.g., IoT sensor networks and physical energy models) [41–43]. While XGBoost typically achieves higher accuracy than RF at the community scale, both models enable efficient identification of energy-saving potentials in urban buildings through morphology-driven feature interpretation [36,44]. However, it still lacks a framework integrating open-source building data with individual energy characteristics for predicting city-scale old residential energy consumption.

To address these research gaps, this study develops a novel approach integrating building energy simulation with machine learning to predict large-scale old residential building energy consumption using multi-source datasets. First, the K-means clustering algorithm was employed to analyze key urban building features (length, width, number of floors, aspect ratio), identifying representative building prototypes. Second, the energy utilization index (EUI) for each prototype was generated using EnergyPlus simulations. Relationships between urban morphological indicators and energy consumption were analyzed. Third, an XGBoost model was developed to predict city-scale energy consumption. Results and their spatial patterns across different scales were visualized. This study provides an efficient framework for analyzing the energy consumption of old urban residential buildings, providing support for urban planning and energy saving strategies.

2. Materials and Methods

The research framework for archetype identification and energy consumption prediction of old residential buildings is shown in Figure 1. Firstly, the data processing process involves collecting multi-source urban building datasets and identifying old residential buildings through house age thresholds and area of interest (AOI). The approximate rectangle method is applied to derive key building parameters, including building length and elevation. Secondly, old residential buildings are categorized into distinct archetypes based on building morphology parameters and clustering algorithms. EnergyPlus, the most widely used energy consumption simulation software, is employed to calculate the energy consumption data of each archetype. This archetype's energy consumption data is then assigned to corresponding residential buildings according to their cluster numbers, thereby providing diverse energy simulation data for the energy consumption prediction model. Thirdly, a building energy prediction model is developed by integrating machine learning algorithms (i.e., XGBoost and RF) with building morphological parameters. City-scale energy consumption of old residential buildings is subsequently estimated using this model. Finally, spatial mapping of urban energy consumption patterns is conducted, and key influencing factors are quantified through interpretable machine learning analysis. This study offers a methodological framework for calculating city-scale energy consumption in old residential buildings, with the results intended to inform targeted retrofit strategies and advance sustainable urban development.

2.1. Study Area and Data Collection

2.1.1. Case Study Area

Guangzhou (120°52′ E to 121°52′ E, 22°40′ N to 23°53′ N) (Figure 2) is situated in south-central Guangdong Province, on the northern edge of China's Pearl River Delta [45]. The city features a marine subtropical monsoon climate, classified as a hot summer and warm winter climate zone [46]. Guangzhou's central urban area encompasses the districts of Yuexiu, Liwan, Tianhe, and Haizhu. This area contains Guangzhou's most densely developed old residential neighborhoods and serves as the city's economic, cultural, and commercial core.

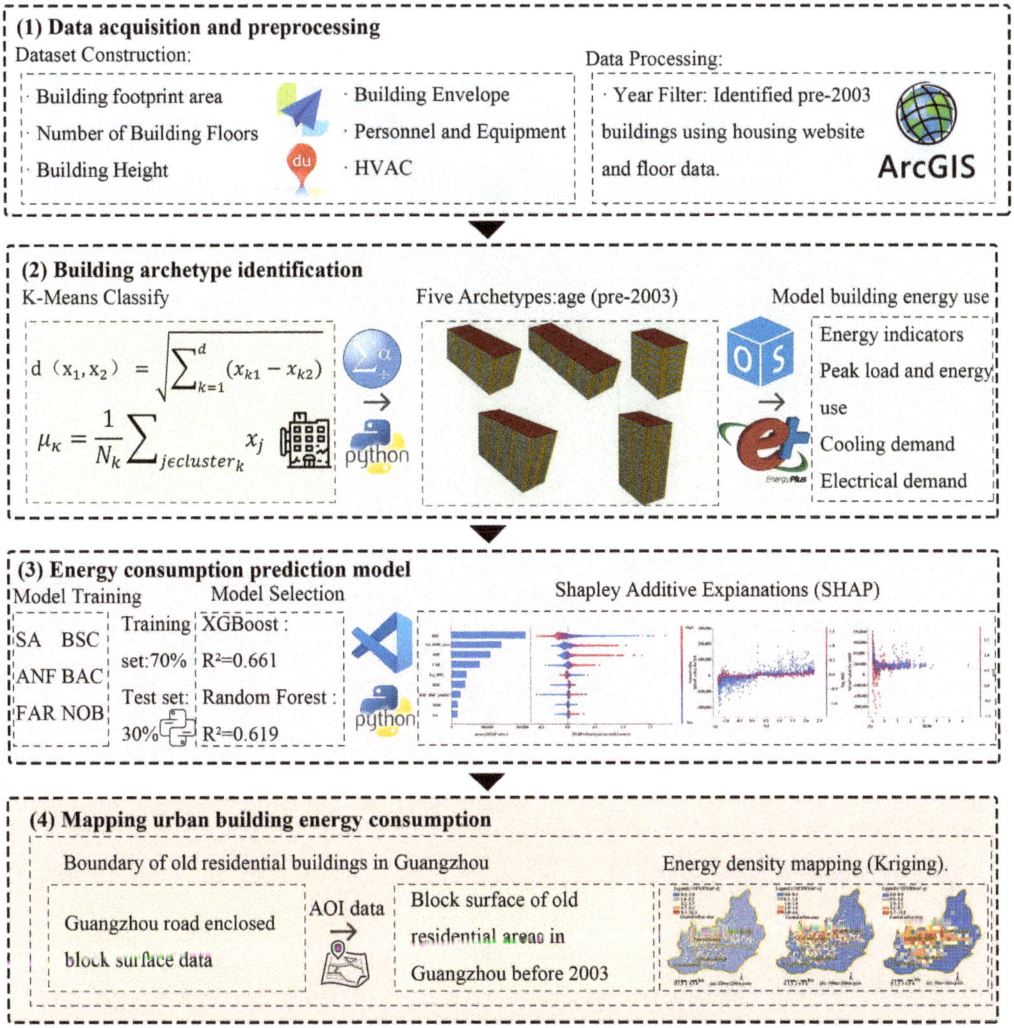

Figure 1. Research framework of energy consumption prediction for old residential buildings.

These neighborhoods primarily consist of courtyard-style housing constructed from the 1950s to the 1990s and commercial housing built during the 1980s and 1990s. During rapid urbanization, numerous old residential buildings emerged in these areas. The layout is relatively regular and simple, with most buildings featuring rectangular designs.

Due to early construction dates and low initial building standards, the thermal performance of building envelope structures in old residential buildings has significantly deteriorated over time [47]. This degradation necessitates substantially higher energy consumption to maintain indoor thermal comfort during Guangzhou's high-temperature, high-humidity summers [48,49]. Consequently, accurate energy consumption assessment in these aging communities represents a critical research priority. This study focuses on Guangzhou's central urban districts (Yuexiu, Liwan, Tianhe, Haizhu) to investigate city-scale energy consumption in old residential communities. These areas contain numerous residential buildings constructed before 2003 that need to conduct energy-saving retrofits.

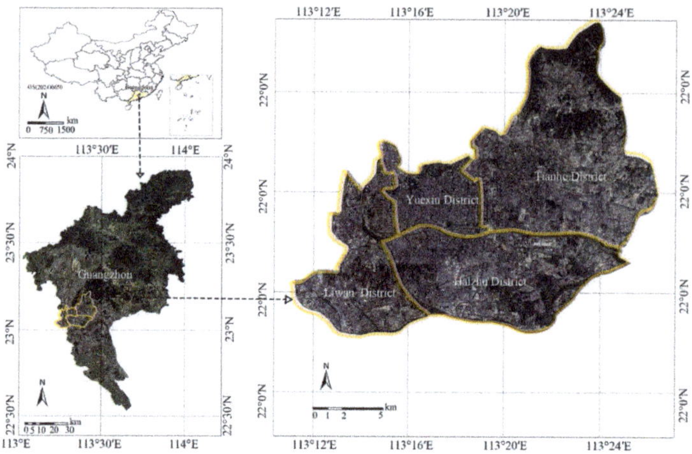

Figure 2. Satellite map of Guangzhou city and old town.

2.1.2. Data Collection and Pre-Processing

Datasets include spatial data, attribute data, and weather data, as shown in Table 1. Building outline data is obtained using Baidu Map API v3.0; road network data is obtained from the OpenStreetMap website. Community age information for developments constructed between 1980 and 2003 was sourced from the Anjuke (www.anjuke.com) platform. Typical Meteorological Year (TMY) and design day data for South China were incorporated into the analysis.

Table 1. Data sources.

Dataset Name	Data Source	Format	Attributes
2023 Guangzhou Building Footprint	Baidu Map (https://map.baidu.com/; accessed 15 May 2023)	Shapefile	Polygon data
Guangzhou Road Network	OpenStreetMap (https://www.openstreetmap.org/)	Shapefile	Line data
Residential Areas with Year Built	Anjuke (https://www.anjuke.com/)	Xlsx	Point data with year tags
Areas of interest	Amap (https://ditu.amap.com/)	Shapefile	Polygon data
Community Coordinates	Baidu Coordinate Picker (https://api.map.baidu.com/lbsapi/getpoint/)	Shapefile	Point data
Meteorological Data	EnergyPlus (https://energyplus.net/)	Epw	Climate parameters

Data preprocessing was conducted as follows. First, building outlines, road networks, and AOIs were processed in ArcGIS. Then, residential construction dates from Anjuke website text records were batch-converted to georeferenced point data via the Baidu Coordinate Picking System. Non-rectangular structures were geometrically standardized through minimum bounding rectangle extraction. Key morphological parameters (length, width, aspect ratio, floor count) were programmatically derived using Python's GeoPandas library (v0.5.1) within the ArcGIS 10.6 environment, generating a structured geodatabase for subsequent analysis.

2.2. Prototype of Old Residential Buildings

2.2.1. K-Means Method

K-means clustering (Figure 3) was selected for its efficiency in grouping buildings by morphology. Building prototypes are established by categorizing structures into multi-story and high-rise types based on morphological features. The classification parameters include building height, long side, short side, aspect ratio, and floor count. A K-means clustering algorithm is employed to group the building dataset. A random sampling heuristic method initializes the cluster centers by selecting them from the dataset. The Euclidean distance metric is used, and the formula is as follows:

$$\mathrm{d}(\mathrm{x}_1, \mathrm{x}_2) = \sqrt{\sum_{k=1}^{d} (x_{k1} - x_{k2})} \tag{1}$$

where $\mathrm{x}_1, \mathrm{x}_2$ are two sample points, d represents feature dimensions (including building height, long side, short side, and aspect ratio), x_{k1}, x_{k2} represent the value of the kth feature.

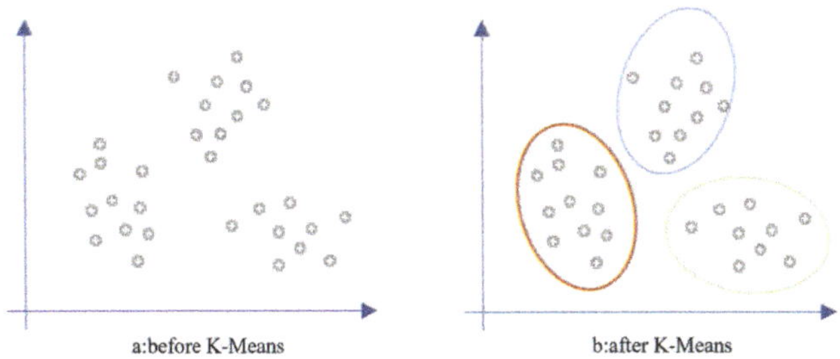

Figure 3. K-Means cluster analysis. (**a**): before K-Means; (**b**): after K-Means.

Samples are assigned to the nearest cluster center according to the minimum distance principle. The new cluster center calculation formula is as follows:

$$\mu_K = \frac{1}{N_k} \sum_{j\epsilon cluster_k} x_j \tag{2}$$

where μ_K is the new center of the k cluster, N_k is the number of samples belonging to the k cluster, x_j is the feature vector of the j sample. The clustering end condition is determined as follows: when the difference between two adjacent cluster centers is less than the set threshold, which means that the clustering criterion function converges, or when the iteration reaches the maximum number of steps, the sample adjustment ends and the final prototype building classification result is obtained.

$$\mathrm{j}(m) = \sum_{j=1}^{k} \sum_{x_i \in c_j} \| x_i - \mu_k \|^2 \tag{3}$$

2.2.2. Parameters of Old Residential Prototypes

The heat transfer coefficients of the exterior wall, roof, and exterior window of the residential building are defined in accordance with the energy-saving design *JGJ 75-2003: Design Standard for Energy Efficiency of Residential Buildings in Hot Summer and Warm Winter Zone;* [50] and previous studies [24]. The exterior walls use clay solid bricks as the primary material, which have low thermal resistance and poor insulation. They also absorb high levels of solar radiation, resulting in increased indoor heat gain during summer. The exterior windows feature aluminum alloy frames, which significantly affect the heat transfer coefficient, causing it to reach 5.0~6.0 $\mathrm{W/(m^2{\cdot}K)}$. The roof is a flat roof with a large-step brick overhead ventilation layer and an expanded perlite insulation layer on top [51]. While this structure primarily optimizes the roof's U-value, its high solar radiation absorption coefficient still leads to excessively high indoor temperatures on the top floor during summer. The definition of thermal parameters of old residential building prototypes is shown in Table 2.

Table 2. Prototype thermal parameter settings.

Parameters	Prototype A	Prototype B	Prototype C	Prototype D	Prototype E
Equipment power density (W/m2)	4.3	4.3	4.3	4.3	4.3
Lighting power density (W/m^2)	7	7	7	7	7
Occupancy (people/m^2)	0.07	0.07	0.07	0.05	0.05
Exterior wall U-value (W/(m^2·K))	2.47	2.47	2.47	1.62	1.62
Roof U-value (W/(m^2·K))	1.8	1.8	1.8	1.66	1.66
Window U-value (W/(m^2·K))	6.4	6.4	6.4	4.99	4.99
Window SHGC	0.85	0.85	0.85	0.85	0.85
Room Air Conditioner SEER	2.7	2.7	2.7	2.7	2.7
Cooling/heating setpoints (°C)	26/16	26/16	26/16	26/16	26/16

China's standard weather data and design day data were selected as input data. The building energy consumption was simulated using the EnergyPlus simulation program. Cooling energy consumption and annual electricity consumption data for different archetypes were then obtained. These energy consumption results for the old residential building archetypes are used to train and test the prediction model.

$$\mathrm{EUI}_i^{\frac{cool}{ele}} = \sum\nolimits_{h\epsilon 8760}\left(\frac{Load_{i,h}^{\frac{cool}{ele}} \times 1}{Floorage_i}\right) \tag{4}$$

where $\mathrm{Load}_{i,h}^{cool/ele}$ denotes the cooling and electricity loads of building i at hour $hFloorage_i$ is the floor area of building i, which is used to calculate the cooling and electricity energy usage intensity per unit area of the whole building.

$$\mathrm{EU}_i^{\frac{cool}{ele}} = \sum\nolimits_{h\in 8760}\left(Load_{i,h}^{\frac{cool}{ele}} \times 1\right) \tag{5}$$

2.3. *Energy Consumption Prediction Model*

This study presents a method for generating urban morphological parameters using open-source data (Figure 4). This method provides an approach to evaluating urban form and its relationship to energy performance at the city scale. The data acquisition and processing phase begins with extracting road network data from OpenStreetMap (OSM), including urban trunk roads, branch roads, and internal roads. These datasets serve as the basis for delineating block boundaries, and the resulting block surfaces are then used for energy analysis. Block feature analysis reveals that larger urban plots frequently incorporate natural landscape elements such as mountains, water bodies, and green spaces, all of which significantly influence the spatial distribution of energy consumption patterns.

To address data gaps in old residential buildings, additional interest surface data was added to construct a more comprehensive block-level dataset. This enhanced dataset captures both the physical characteristics of urban blocks and their functional attributes, enabling accurate energy performance assessments across diverse urban blocks. For spatial visualization, the study employs grid-based techniques to map energy consumption across Guangzhou's central urban area. This gridded representation not only facilitates intuitive interpretation of energy use patterns but also supports comparative analysis between different urban blocks. Such visualization maps serve as valuable tools for urban planning decisions and targeted energy management strategies, particularly in dense urban environments.

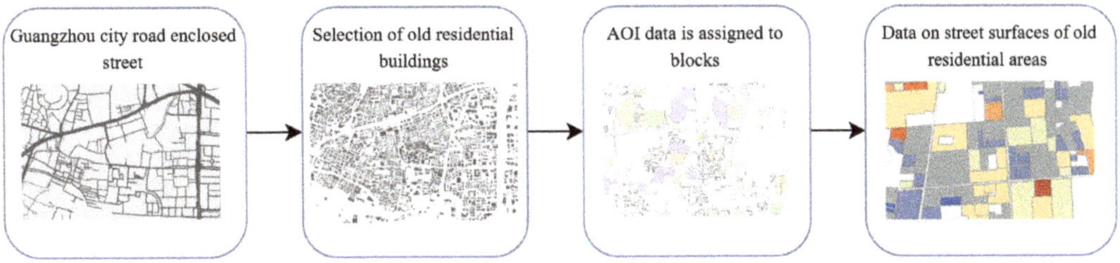

Figure 4. The process of generating the street surface of old residential areas.

The boundary definitions and building form variables of old residential areas are shown in Table 3. These variables serve as inputs for both the XGBoost model and the RF model. Hyperparameter tuning was applied to optimize model performance, with the best parameter combination selected based on evaluation via K-Fold Cross-Validation. The energy consumption prediction accuracy of the two models is compared, and the superior one is chosen for energy analysis of urban old residential buildings.

Table 3. Parameter calculation formula.

Variable	Definition	Formula	Reference
Building Serial Number (SN)	Unique identifier of a building, used to distinguish and mark different buildings.		
Site Area (SA)	Refers to the area of the plot of land on which the building belongs.		
Average number of building floors (ANF)	The average number of floors in a building's total height, usually expressed as an integer or decimal, used to estimate the building's mass.	$ANF = \frac{FAR}{BCR}$	[51]
Number of Buildings (NoB)	The number of buildings within the site calculation unit	NoB = Count of Buildings in the Area	
Building Coverage Ratio (BCR)	The ratio of building area to total land area.	$BCR = \frac{building\ area}{site\ area}$	[52]
Floor Area Ratio (FAR)	The ratio of the total building area in a region to the total area of the region.	$FAR = \frac{building\ floor\ area}{site\ area}$	[53]
Building Shape Coefficient (BSC)	Used to measure the complexity of a building's outline; the ratio of the building's surface area to the unit building volume.	$BSC = \frac{\sum_i^n S_i}{\sum_i^n V_i}$	[54]

Shapley Additive Explanations (SHAP) was utilized to analyze the optimal model's performance, assessing the contribution of six features to the model output. SHAP, derived from cooperative game theory, measures the marginal contribution of features to the model's output by calculating Shapley values. This approach constructs an additive interpretation model where all features are considered contributors. Each feature's contribution to the predicted value is quantified through the Shapley value, ultimately determining the feature importance ranking of the optimal model.

The study selected parameters such as R^2 and Mean Squared Error (MSE) as model performance comparison parameters. R^2 measures the proportion of variance explained by the model for the target variable Energy. The R^2 value range is between 0 and 1. In terms of explanatory power, the closer R^2 is to 1, the better the model fits the data. MSE is the average of the squares of the difference between the predicted value and the true value. The calculation formula is as follows:

$$R^2 = 1 - \frac{\sum_{i=1}^{n}(y_i - \hat{y}_i)^2}{\sum_{i=1}^{n}(y_i - \overline{y})^2} \tag{6}$$

where n represents the number of data points for calculating R^2, y_i is the true value of the i sample, $\hat{y}_i$ represents the predicted value of the ith sample, and $\overline{y}$ represents the average of all true values.

$$\mathrm{MSE} = \frac{1}{n}\sum_{i=1}^{n}(y_i - \hat{y}_i)^2 \tag{7}$$

where n represents the number of data points, y_i is the true value of the i sample, and $\hat{y}_i$ represents the predicted value of the i sample.

2.4. Mapping Energy Consumption

The energy usage of old residential buildings at the city scale is obtained by summarizing the energy usage of each building. The EUI of old residential buildings and its spatial distribution differences were visually analyzed using a heat map. Three grid units (i.e., 220 m × 220 m, 500 m × 500 m, and 1 km × 1 km) were selected to establish a product calculation model based on building archetype area and EUI. Color gradient coding (bright colors indicating low energy consumption, dark colors indicating high energy consumption) was used for multi-scale energy consumption mapping. The 1 km grid reflects the overall energy consumption level of the urban area, the 500 m grid identifies high-consumption clustered areas, and the 220 m grid analyzes block-level distribution characteristics.

3. Results

3.1. Energy Consumption of Prototypes

Satellite images of old residential areas in various districts of Guangzhou from the 1980s to the 1990s were selected to verify the screening results, and the accuracy rate of old residential building recognition reached 90%. Due to the absence of a small number of building outlines and errors in the number of floors in the original building outline data, a small number of old residential buildings could not be identified.

Regarding prototype clustering, the K-means model demonstrated a significant processing effect when the overall data volume exceeded 20,000 samples. The contour coefficient derived from the algorithm was at a medium level, suggesting the presence of a clustering structure. The Davies–Bouldin index of 0.798 indicated substantial distances between clusters while maintaining close intra-cluster distances. The prototype categories and quantity statistics of old residential buildings are shown in Table 4 and Figure 5.

In practice, energy consumption measurements for old residential buildings are unavailable. Thus, the energy consumption results of the old residential building archetypes were validated against previous studies. Cooling energy and EUI results of different prototypes are shown in Figure 6. The average specific power consumption for civil buildings in Guangdong Province in 2012 was 72.13 kWh/(m^2·a) [55]. Since civil buildings encompass both residential and public structures, this average is notably higher than the energy consumption values reported in the study. The discrepancies between Prototypes B and C in the study's results were 10–11%, while the cooling energy consumption of Prototypes A, E, and F closely aligned with the simulated air conditioning energy consumption. Pre-2001 energy consumption figures for old mid-rise residential buildings in hot summer and warm winter regions were 52 kWh/ m^2, and for old high-rise buildings, 38 kWh/m^2 [24]. The mid-rise buildings exhibited a 22% difference compared to the results of this study.

Table 4. Prototype categories of old residential buildings.

Prototype Category	Number	Building Aspect Ratio	Length × Width	Graphic
Prototype A	9708	3.10	42 × 17	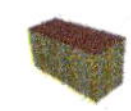
Prototype B	2394	4.01	78 × 26	
Prototype C	16,114	1.94	21 × 12	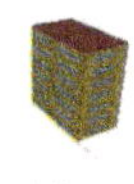
Prototype D	11,057	2.09	30 × 17	
Prototype E	3575	3.21	73 × 27	

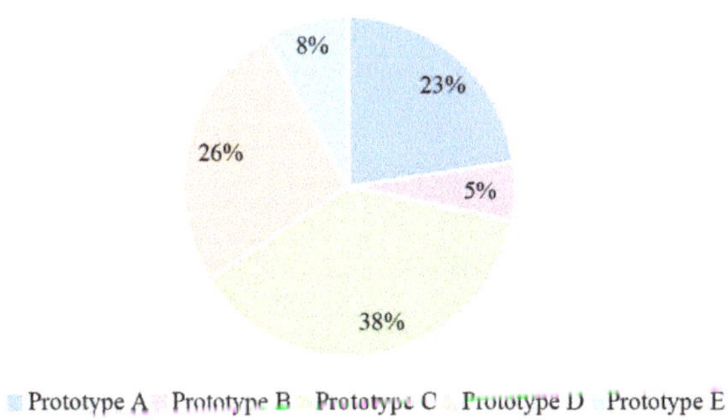

Figure 5. Proportion of prototypes of old residential buildings in Guangzhou.

3.2. Prediction of Building Energy Consumption at City Scale

3.2.1. Comparison of Different Prediction Models

Machine learning prediction analysis uses building-scale and city-scale cooling energy consumption and electricity usage data. By comparing the performance of XGBoost and RF on the test set, it is found that XGBoost's R^2 (0.667) is slightly higher than RF (0.619), indicating that the Boosting method has stronger nonlinear fitting ability on this dataset. The higher unit value of the predicted energy consumption value leads to a larger MSE indicator value, 1.0×10^{11} for XGBoost and 1.1×10^{11} for RF. The model performance R^2 of the XGBoost model in predicting the complex relationship between urban spatial morphology and land surface temperature in the plain city of Chengdu and the plateau city of Lhasa is 0.515 and 0.429, respectively [56]. Similar studies have shown an R^2 of 0.685 in the test set when predicting cooling loads in large urban communities [34]. The R^2 in the study of the relationship between urban form and urban building energy consumption is 0.4854 [57]. In summary, the R^2 of the XGBoost model in prediction is 0.667, which is within a reasonable range.

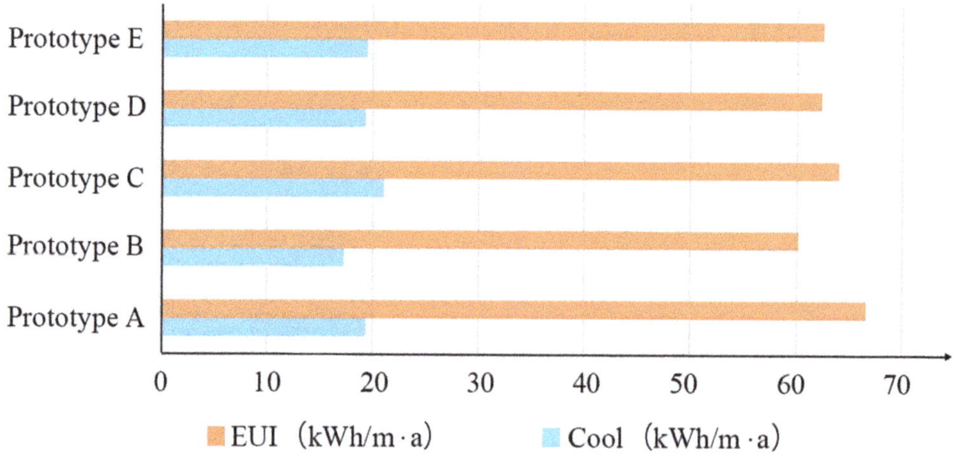

Figure 6. Annual energy consumption and cooling energy consumption of different prototypes.

This study uses Random Forest as an alternative model for cross-verification to evaluate the robustness of the XGBoost model. The superparameters of the random forest (including the number of decision trees, the minimum sample number of leaf nodes, and the maximum number of features) are optimized by the RandomizedSearchCV method, and the model is selected based on the 3-fold cross-verification and mean square error minimization target. The performance of the optimal random forest model on the test set is compared with that of the XGBoost model. The results show that the two have a high consistency in the order of feature importance (Spearman correlation coefficient = 0.82, $p < 0.01$), indicating that the model is stable in the identification of key features. In addition, the robustness of the model was verified through data segmentation experiments of different random seeds (42, 123, 456, 789), and the R^2 fluctuation range was controlled within ± 0.03, confirming the reliability of the model's prediction performance.

3.2.2. Analysis of Influencing Factors

Based on the urban morphological parameters and building energy consumption datasets generated by residential building energy consumption prototypes and blocks. This study used the SHAP method to quantify the marginal contribution of each feature to the prediction results, and constructed a building energy consumption prediction framework that integrates ensemble learning and interpretable analysis for energy consumption prediction analysis of old residential buildings in Guangzhou.

The feature correlation matrix (Figure 7) shows the linear correlation between the features in the dataset: SA and ANF: highly negative correlation (about −0.99), indicating that when SA increases, ANF almost always decreases. BSC and energy: the positive correlation is 0.52, indicating that BSC has a positive impact on the change in energy. BAC and energy: high positive correlation (about 0.79), which means that BAC is an important factor affecting energy. In terms of correlation with the energy of the target variable: BAC and BSC: there is a strong positive correlation with energy. ANF_BSC_product(ANF × BSC) and log_BSC(Logarithmic transformation of building shape coefficients): there is a moderate correlation with energy, indicating that housing project characteristics such as BAC and BSC can help improve the prediction effect.

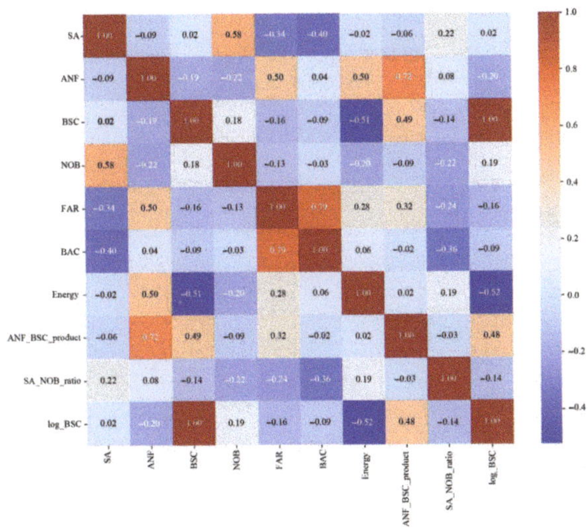

Figure 7. Feature correlation matrix for building energy prediction.

The feature importance plots in Figure 8 show that BSC has a higher impact, while SA has the lowest impact. The SHAP values (impact on model output) show that features BSC, ANF, and FAR have the greatest impact on model output, and they significantly positively promote the output results.

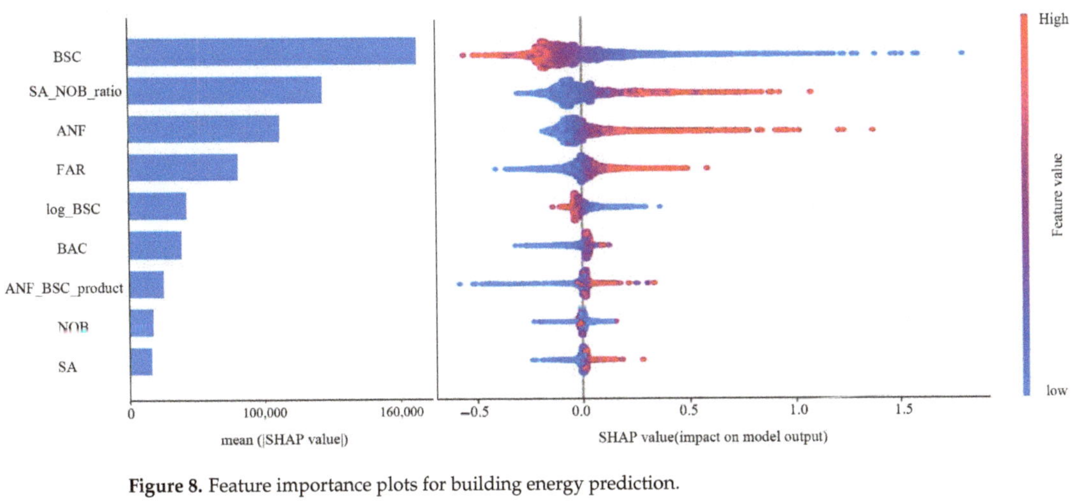

Figure 8. Feature importance plots for building energy prediction.

Figure 9 reveals that building morphology parameters particularly ANF, FAR, and BSC are the dominant positive drivers of energy consumption, contributing SHAP values of +414,305.15, +393,685.13, and +268,099.79, respectively. Conversely, Surface Area (SA) and Number of Buildings (NoB) emerge as negative factors, reducing energy consumption predictions by −8678.71 and −5184.37, respectively. This directional influence is visually evident in the distribution of feature impacts, where higher values of positive drivers (red data points) consistently push predictions above the baseline output (E[f(x)] = 402,997), while higher values of negative factors (blue) correlate with reduced consumption estimates.

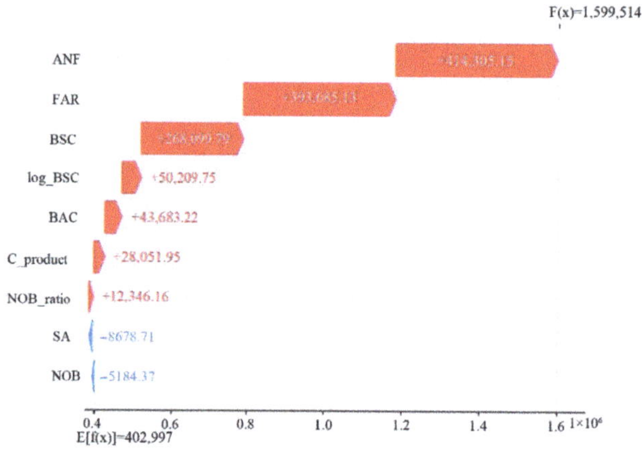

Figure 9. SHAP decision plot for building energy prediction.

Figure 10 delineates the SHAP value distributions for six critical features (a–f) in an XGBoost model, revealing distinct characteristic patterns. In Figure 10a, the SA feature demonstrates a bimodal distribution of SHAP values (range: −1 to 1), with medium-value samples (blue) exhibiting predominantly negative contributions, while high-value samples (red) show heterogeneous positive effects. Figure 10b discloses a pronounced clustering of high NoB values (red) at SHAP $\approx$ 0.5, indicating a systematic positive impact beyond specific thresholds. The FAR feature in Figure 10c exhibits a weak linear tendency, where low-value samples (blue) concentrate in negative SHAP space, contrasting with high-value samples (red) in positive regions. Figure 10d manifests a polarized pattern for BAC: low values drive strong negative contributions (SHAP < −0.5), whereas high values correlate with robust positive effects (SHAP > 1). The BSC feature in Figure 10e displays minimal dispersion (range: ±0.25 SHAP) without clear value-gradient dependency, suggesting stable predictive influence. Notably, Figure 10f reveals an abrupt transition for ANF—low values yield negative contributions, but marginal increments trigger a phase shift to sustained positive SHAP values (>0.6), evidencing a critical nonlinear boundary effect.

3.3. Analysis of Old Residential Building Energy Consumption

3.3.1. Energy Consumption Analysis at Urban Scale

The spatial distribution of energy consumption in old residential areas reveals a concentration in peri-urban areas surrounding Guangzhou's historic core. As shown in Figure 11, At the 1 km grid-scale analysis, Huadu Central District demonstrates significantly higher energy intensity (27.1–94.3 × 10^5 kWh/m^2), contrasting with lower intensity values (27.1 × 10^5 kWh/m^2) observed in Zengcheng and Conghua urban centers. The least energy-intensive areas, corresponding to minimal old housing stock, are primarily located in Nansha District and Huangpu District. Within the four central historic districts, energy consumption displays high spatial clustering with intensities predominantly below 9.69 × 10^5 kWh/m^2, necessitating finer-scale analysis (e.g., 500 m grid) to delineate precise energy intensity zones.

3.3.2. Energy Consumption Differences Under Different Grids

In Figure 12, this study analyzed urban energy consumption patterns in Guangzhou's central districts—Liwan, Yuexiu, and Haizhu—employing a multi-scale grid approach (220 × 220 m, 500 × 500 m, and 1000 × 1000 m). Total energy consumption in old residential areas was estimated by aggregating EUI values with corresponding floor areas of prototype buildings. At the 1 km × 1 km scale, high-consumption zones (8.7–15.8 × 10^7 kWh/m^2·a) clustered predominantly in Yuexiu District, moderate-consumption areas (2.6–8.7 × 10^7 kWh/m^2·a) concentrated in Tianhe and Haizhu, while Liwan District showed low consumption (0.9 × 10^7 kWh/m^2·a). The 500 m × 500 m resolution revealed finer distributions: Yuexiu's sub-districts exhibited moderate consumption (3.0–4.6 × 10^6 kWh/m^2·a), Tianhe contained more low-consumption areas (0.3–1.8 × 10^6 kWh/m^2·a) than Haizhu, and minimal consumption zones (0.3 × 10^6 kWh/m^2·a) dominated Liwan. Neighborhood-level analysis at 220 m × 220 m resolution identified peak intensity clusters (5.7–18.3 × 10^5 kWh/m^2·a) in Yuexiu's Jianshe Sub-district, effectively capturing block-level variations in old residential areas.

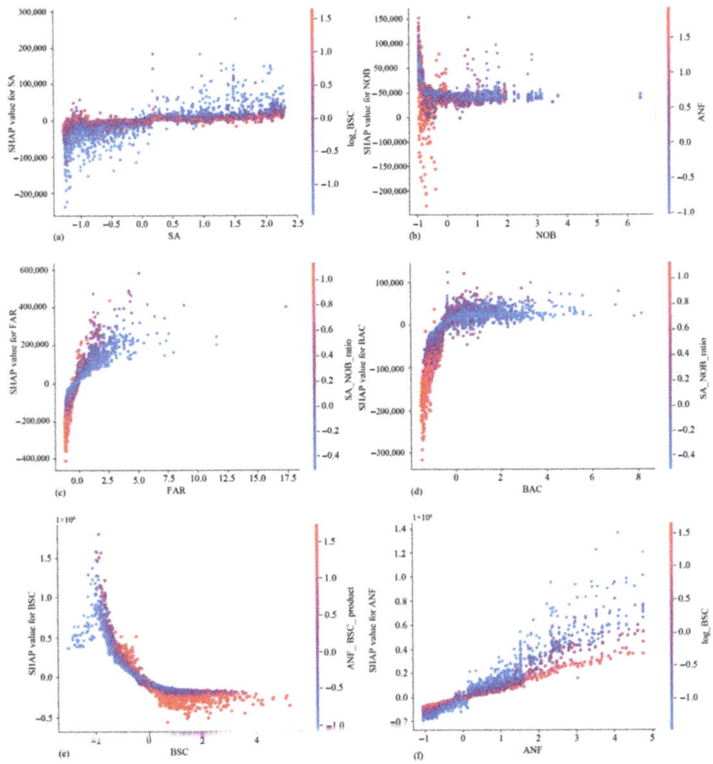

Figure 10. SHAP interaction summary plots: (**a**) SHAP value for SA; (**b**) SHAP value for NoB; (**c**) SHAP value for FAR; (**d**) SHAP value for BAC; (**e**) SHAP value for BSC; (**f**) SHAP value for ANF.

3.3.3. Energy Consumption of Different Prototypes

Two types of old residential areas, namely Suihua New Village in the 1980s and Liuyun Community in the 1990s, were selected, and the number and distribution of their building prototypes were counted (A/C/D types accounted for more than 80%). The impact of envelope structure defects on energy consumption was simulated with Lingnan climate parameters, revealing the common laws of energy consumption of building prototypes from different eras. The study selected two old residential communities of different ages, Liuyun Community and Suihua New Village in Guangzhou, to analyze the distribution and energy consumption characteristics of different residential prototypes.

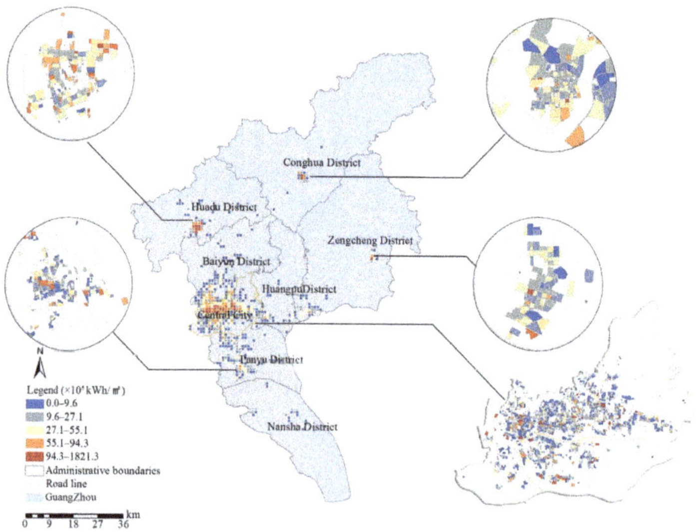

Figure 11. Visualization of energy consumption in old residential areas in Guangzhou.

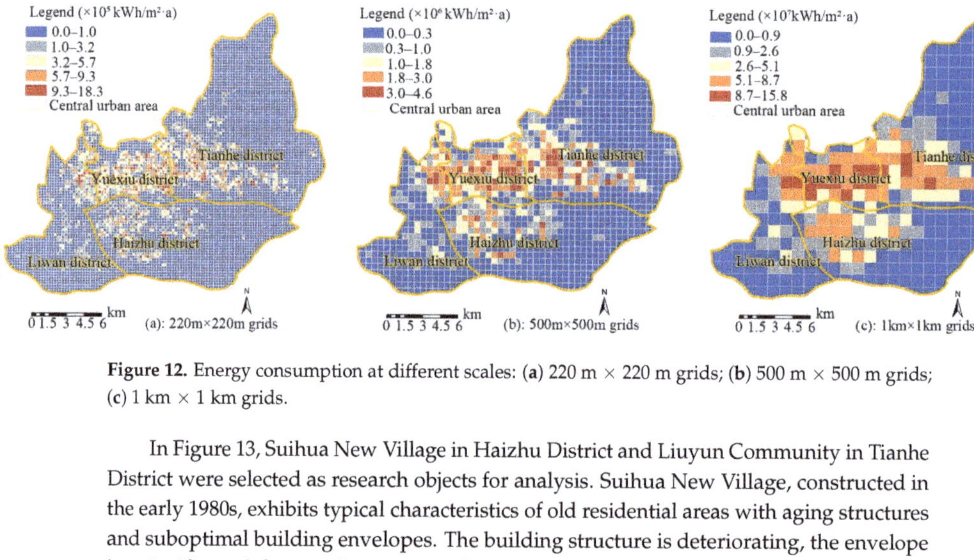

Figure 12. Energy consumption at different scales: (**a**) 220 m × 220 m grids; (**b**) 500 m × 500 m grids; (**c**) 1 km × 1 km grids.

In Figure 13, Suihua New Village in Haizhu District and Liuyun Community in Tianhe District were selected as research objects for analysis. Suihua New Village, constructed in the early 1980s, exhibits typical characteristics of old residential areas with aging structures and suboptimal building envelopes. The building structure is deteriorating, the envelope has significant defects, and supporting facilities are relatively outdated, leading to elevated energy consumption in the high-temperature and high-humidity Lingnan climate. Liuyun Community, built in the early 1990s, represents a typical open, pure residential community developed by private investors. The buildings' exteriors are in poor condition, and their energy consumption patterns demonstrate strong universality. Furthermore, since both communities underwent certain renovation and transformation processes during urban development, the energy consumption simulation parameters were not specifically differentiated for each era. The buildings in these two communities predominantly correspond to Prototype A, C, and D categories, reflecting commonalities in the prototype characteristics of older communities.

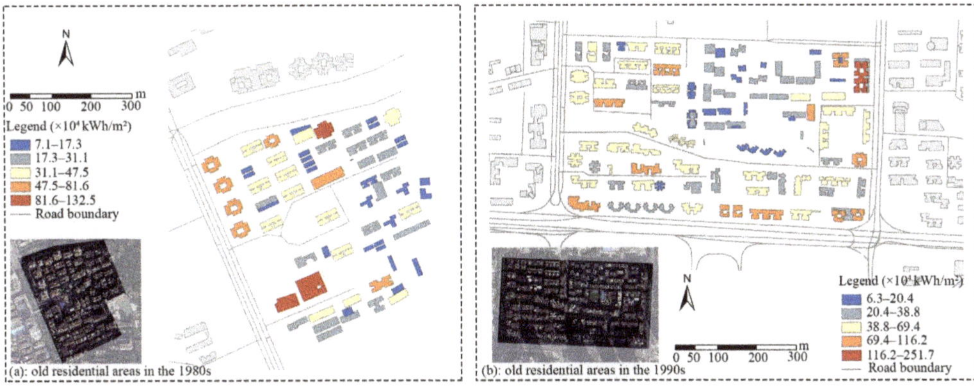

Figure 13. Spatial distribution of building energy consumption: (**a**) old residential areas from the 1980s; (**b**) old residential areas from the 1990s.

From the perspective of urban renewal, the energy consumption conditions of the old residential area are introduced into the urban renovation and renewal guidelines to guide the energy-saving and low-carbon development of old communities. As shown in Figure 14, there are various strategies for the renovation of urban residential buildings. At block level/building scale, strengthen the promotion of energy-saving and low-carbon measures in communities, open up public activity spaces in neighborhoods, establish community ventilation micro-channels, enhance residents' awareness of low-carbon and energy-saving, strengthen roof and exterior wall structures, and use more environmentally friendly and energy-saving doors and windows.

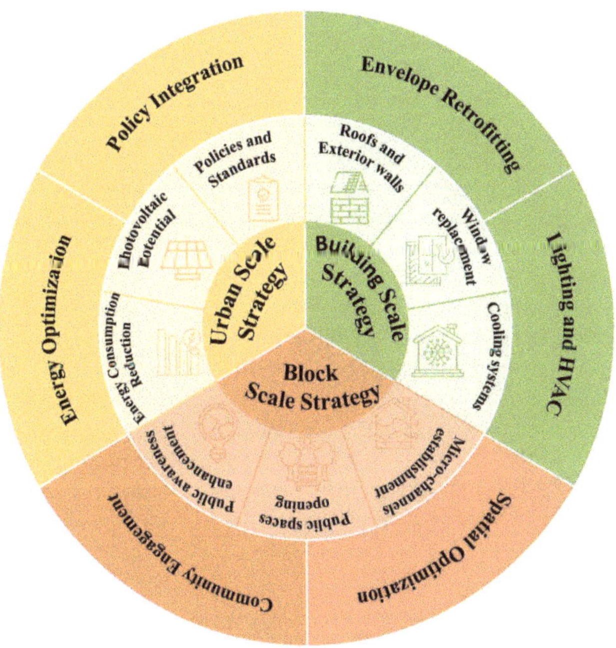

Figure 14. Multi-scale energy-saving and low-carbon renewal measures for old residential areas.

4. Discussion

The old residential prototype building is determined based on the building length and width, aspect ratio, and number of floors, and OpenStudio is used to efficiently simulate the energy consumption of the old residential prototype building. Compared with the energy consumption of old residential buildings in different climate zones in the previous literature, the simulation results of the old residential prototype building are within a reasonable range, and the actual building energy consumption data is needed to further calibrate the building energy consumption.

Due to the lack of actual measured energy consumption data of an old residential building, the study refers to the data in previous research and relevant documents for verification, and uses the data in the Construction Yearbook to verify the results. According to the 2015 Guangdong Construction Yearbook, the average unit area power consumption of civil buildings in the 2012 Guangdong Provincial Energy Audit is 72.13 kilowatt-hours/(m^2·a) [55]. The energy audit results include residential buildings and public buildings, so they are significantly higher than the energy consumption values obtained in the study. The energy consumption simulation results of the previous study using the Guangzhou Asian Games City residential building as an example showed that the air conditioning cooling energy consumption was 19.3 kWh/m^2. This value differs from the cooling energy consumption of Prototype B and Prototype C by 10–11%, while the cooling energy consumption of Prototype A, Prototype E, and Prototype F is very close to the research results [58]. The annual electricity EUI of the prototype building of mid-rise residential buildings in Changsha, a hot summer and cold winter region, is 50.2 kWh/m^2 [24], which is 16.4% to 24.5% different from the annual electricity EUI of the prototype building obtained in this study. The cooling energy consumption demand is higher in the hot summer and warm winter region throughout the year. Therefore, the building energy data is within a reasonable error range.

Urban morphological parameters and energy consumption data are used as inputs for machine learning algorithms. The XGBoost and RF models are then employed to estimate the energy consumption of old residential buildings. XGBoost achieves higher prediction accuracy than the RF model. The energy consumption of old urban areas is analyzed at three different scales (220 m, 500 m, and 1 km), providing a reference for further identifying energy consumption patterns and informing energy planning.

This study still has some limitations. A typical meteorological year's data was selected as input for the building energy model in this study. However, extreme high temperatures, the urban heat island effect, as well as usage habits and population density can significantly impact building energy consumption results [59–61]. Future work should consider these impacts and further improve the energy performance simulation of old residential buildings [62]. Meanwhile, the predictive accuracy of the XGBoost model remains limited. Future research should focus on enhancing accuracy through optimization of the prototype datasets. In addition, the application of the proposed residential building energy prediction model in other cities needs to be validated in the future. The model will require adaptive adjustments based on the actual meteorological parameters of the study area and the structural characteristics of old residential buildings.

5. Conclusions

This study developed a novel approach that integrates building energy simulation and machine learning to predict city-scale energy use in old residential buildings, utilizing multi-source datasets. It extends building-scale simulations to estimate energy consumption at the city scale. The main conclusions include:

- Pre-2003 residential buildings were identified from 706,188 building outlines in Guangzhou using AOI data and internet housing data. This identification method was validated with 90% accuracy, yielding a total of 31,209 confirmed pre-2003 residential buildings;
- Five representative prototypes exhibited cooling energy use (17.32–21.05 kWh/m^2) and annual electricity EUI (60.10–66.53 kWh/m^2), scalable to city-scale energy assessment for low-carbon planning;
- A prediction model establishing the correlation between urban morphological factors and energy consumption was developed. The reliability of the XGBoost model was confirmed through cross-validation and its predictive accuracy for energy consumption. Model performance demonstrated that the XGBoost algorithm (R^2 = 0.667) outperformed the RF model. Furthermore, a strong positive correlation (r = 0.79) was identified between BSC and the energy consumption of old residential buildings;
- Spatial analysis of Guangzhou's old residential buildings revealed distinct energy consumption patterns: Huadu Central District exhibited peak intensity (27.1–94.3 $\times$ 10^5 kWh/m^2), while Nansha and Huangpu showed relatively lower consumption levels. Multi-scale grid analysis (ranging from 220 m to 1 km) identified Yuexiu as the highest energy consumption zone, with peak values reaching up to 18.3 $\times$ 10^5 kWh/m^2·a at 220 m resolution. Prototype energy simulations of 1980s–1990s communities indicated that building envelope was the key inefficiency factor.

This study establishes an efficient framework for analyzing the energy consumption of urban old residential buildings. This framework facilitates the development of urban planning strategies and energy-saving policies. Potential applications include: (1) At city scale, it can support the design of policies, standards, and guidelines to promote energy efficiency in urban renewal projects. (2) At block scale, it can enhance public awareness and transform public spaces within high-energy-consumption communities to encourage energy-saving and low-carbon practices among residents, fostering both promoted and spontaneous energy-conserving behaviors. (3) At building scale, it can guide the retrofitting of building envelope components (e.g., roofs, windows, exterior walls) to reduce the energy consumption of individual structures.

Author Contributions: Conceptualization, C.F. and R.L.; methodology, C.F., R.L., and Y.L.; formal analysis, C.F., R.L., and Y.L.; resources, Y.L.; data curation, C.F. and R.L.; writing—original draft preparation, R.L. and C.F.; writing—review and editing, Y.L.; visualization, supervision, C.F. and Y.L.; funding acquisition, C.F. All authors have read and agreed to the published version of the manuscript.

Funding: The research was supported by Guangdong Basic and Applied Basic Research Foundation (No. 2023A1515012138; 2025A1515012875), Guangzhou Science and Technology Project (No. 2024A04J3355).

Data Availability Statement: The datasets generated and analyzed during the current study are available from the corresponding author on reasonable request.

Conflicts of Interest: The authors declare that they have no known competing financial interest or personal relationships that could have appeared to influence the work reported in this paper.

Abbreviations

XGBoost	eXtreme Gradient Boosting
RF	Random Forest
SHAP	Shapley Additive Explanations
BSC	Building Shape Factor
EUI	Energy Use Intensity
SN	Building Serial Number
SA	Site Area
SHGC	Solar Heat Gain Coefficient
SEER	Seasonal Energy Efficiency Ratio
ANF	Average number of building floors
NoB	Number of Buildings
BCR	Building Coverage Ratio
FAR	Floor Area Ratio
MSE	Mean Squared Error

References

1. WCR PR Chinese Press Release. wuf.unhabitat. 2022. Available online: https://wuf.unhabitat.org/sites/default/files/2022-06/files/WCR-PR-Chinese-press-release-29-06-2022.pdf (accessed on 20 June 2025).
2. Zou, B.; Fan, C.; Wang, M.; Li, J.; Zhou, X.; Liao, Y. How Do Urban-Rural and Regional Summer Heat Exposures Evolve? A Case Study of 301 Cities in China from 2000 to 2020. *Build. Environ.* **2025**, *271*, 112584. [CrossRef]
3. Wang, Z.; Hong, Y.; Huang, L.; Zheng, M.; Yuan, H.; Zeng, R. A Comprehensive Review and Future Research Directions of Ensemble Learning Models for Predicting Building Energy Consumption. *Energy Build.* **2025**, *335*, 115589. [CrossRef]
4. *Statistical Communiqué of the People's Republic of China on National Economic and Social Development in 2024*; National Bureau of Statistics: Beijing, China, 2025.
5. *Annual Research Report on China Building Energy Efficiency 2025 (Urban Residential Buildings Special Topic)*; Tsinghua University Building Energy Research Center: Beijing, China, 2025.
6. *Action Plan for Carbon Dioxide Peaking Before 2030*; Department of Resource Conservation and Environmental Protection: Beijing, China, 2021.
7. Peng, Z.; Sun, Q.; Li, P.; Sun, F.; Ren, S.; Guan, R. The Environmental Impact of the Entire Renovation Process of Urban Aged Residential Buildings in China. *IJBPA* **2024**. [CrossRef]
8. Dahlström, L.; Broström, T.; Widén, J. Advancing Urban Building Energy Modelling through New Model Components and Applications: A Review. *Energy Build.* **2022**, *266*, 112099. [CrossRef]
9. Ferrando, M.; Causone, F.; Hong, T.; Chen, Y. Urban Building Energy Modeling (UBEM) Tools: A State-of-the-Art Review of Bottom-up Physics-Based Approaches. *Sustain. Cities Soc.* **2020**, *62*, 102408. [CrossRef]
10. Hong, T.; Chen, Y.; Luo, X.; Luo, N.; Lee, S.H. Ten Questions on Urban Building Energy Modeling. *Build. Environ.* **2020**, *168*, 106508. [CrossRef]
11. Abbasabadi, N.; Ashayeri, M. Urban Energy Use Modeling Methods and Tools: A Review and an Outlook. *Build. Environ.* **2019**, *161*, 106270. [CrossRef]
12. Gan, L.; Liu, Y.; Shi, Q.; Cai, W.; Ren, H. Regional Inequality in the Carbon Emission Intensity of Public Buildings in China. *Build. Environ.* **2022**, *225*, 109657. [CrossRef]
13. Johari, F.; Peronato, G.; Sadeghian, P.; Zhao, X.; Widén, J. Urban Building Energy Modeling: State of the Art and Future Prospects. *Renew. Sustain. Energy Rev.* **2020**, *128*, 109902. [CrossRef]
14. Koral Iseri, O.; Duran, A.; Canlı, I.; Meral Akgul, C.; Kalkan, S.; Gursel Dino, I. A Method for Zone-Level Urban Building Energy Modeling in Data-Scarce Built Environments. *Energy Build.* **2025**, *337*, 115620. [CrossRef]
15. Yang, J.; Zhang, Q.; Peng, C.; Chen, Y. AutoBPS-Prototype: A Web-Based Toolkit to Automatically Generate Prototype Building Energy Models with Customizable Efficiency Values in China. *Energy Build.* **2024**, *305*, 113880. [CrossRef]
16. Oraiopoulos, A.; Hsieh, S.; Schlueter, A. Energy Futures of Representative Swiss Communities under the Influence of Urban Development, Building Retrofit, and Climate Change. *Sustain. Cities Soc.* **2023**, *91*, 104437. [CrossRef]
17. Cerezo, C.; Sokol, J.; AlKhaled, S.; Reinhart, C.; Al-Mumin, A.; Hajiah, A. Comparison of Four Building Archetype Characterization Methods in Urban Building Energy Modeling (UBEM): A Residential Case Study in Kuwait City. *Energy Build.* **2017**, *154*, 321–334. [CrossRef]
18. Shen, P.; Wang, H. Archetype Building Energy Modeling Approaches and Applications: A Review. *Renew. Sustain. Energy Rev.* **2024**, *199*, 114478. [CrossRef]

19. Henninger, R.H.; Witte, M.J.; Crawley, D.B. Analytical and Comparative Testing of EnergyPlus Using IEA HVAC BESTEST E100–E200 Test Suite. *Energy Build.* **2004**, *36*, 855–863. [CrossRef]
20. Yan, D.; Zhou, X.; An, J.; Kang, X.; Bu, F.; Chen, Y.; Pan, Y.; Gao, Y.; Zhang, Q.; Zhou, H.; et al. DeST 3.0: A New-Generation Building Performance Simulation Platform. *Build. Simul.* **2022**, *15*, 1849–1868. [CrossRef]
21. Carnieletto, L.; Ferrando, M.; Teso, L.; Sun, K.; Zhang, W.; Causone, F.; Romagnoni, P.; Zarrella, A.; Hong, T. Italian Prototype Building Models for Urban Scale Building Performance Simulation. *Build. Environ.* **2021**, *192*, 107590. [CrossRef]
22. Liu, M.; Gou, Z. A Regional Domestic Energy Consumption Model Based on LoD1 to Assess Energy-Saving Potential. *Adv. Eng. Inform.* **2025**, *65*, 103247. [CrossRef]
23. Deng, Z.; Chen, Y.; Pan, X.; Peng, Z.; Yang, J. Integrating GIS-Based Point of Interest and Community Boundary Datasets for Urban Building Energy Modeling. *Energies* **2021**, *14*, 1049. [CrossRef]
24. Deng, Z.; Chen, Y.; Yang, J.; Chen, Z. Archetype Identification and Urban Building Energy Modeling for City-Scale Buildings Based on GIS Datasets. *Build. Simul.* **2022**, *15*, 1547–1559. [CrossRef]
25. Song, C.; Deng, Z.; Zhao, W.; Yuan, Y.; Liu, M.; Xu, S.; Chen, Y. Developing Urban Building Energy Models for Shanghai City with Multi-Source Open Data. *Sustain. Cities Soc.* **2024**, *106*, 105425. [CrossRef]
26. An, J.; Wu, Y.; Gui, C.; Yan, D. Chinese Prototype Building Models for Simulating the Energy Performance of the Nationwide Building Stock. *Build. Simul.* **2023**, *16*, 1559–1582. [CrossRef]
27. Peng, H.; Li, M.; Lou, S.; He, M.; Huang, Y.; Wen, L. Investigation on Spatial Distribution and Thermal Properties of Typical Residential Buildings in South China's Pearl River Delta. *Energy Build.* **2020**, *206*, 109555. [CrossRef]
28. Xia, D.; Wu, Z.; Zou, Y.; Chen, R.; Lou, S. Developing a Bottom-up Approach to Assess Energy Challenges in Urban Residential Buildings of China. *Front. Archit. Res.* **2025**, S209526352500041X. [CrossRef]
29. Yu, H.; Wang, M.; Lin, X.; Guo, H.; Liu, H.; Zhao, Y.; Wang, H.; Li, C.; Jing, R. Prioritizing Urban Planning Factors on Community Energy Performance Based on GIS-Informed Building Energy Modeling. *Energy Build.* **2021**, *249*, 111191. [CrossRef]
30. Alasmar, R.; Schwartz, Y.; Burman, E. Developing a Housing Stock Model for Evaluating Energy Performance: The Case of Jordan. *Energy Build.* **2024**, *308*, 114010. [CrossRef]
31. Li, X.; Yao, R.; Liu, M.; Costanzo, V.; Yu, W.; Wang, W.; Short, A.; Li, B. Developing Urban Residential Reference Buildings Using Clustering Analysis of Satellite Images. *Energy Build.* **2018**, *169*, 417–429. [CrossRef]
32. Bourdeau, M.; Zhai, X.Q.; Nefzaoui, E.; Guo, X.; Chatellier, P. Modeling and Forecasting Building Energy Consumption: A Review of Data-Driven Techniques. *Sustain. Cities Soc.* **2019**, *48*, 101533. [CrossRef]
33. Sheng, Y.; Arbabi, H.; Ward, W.O.C.; Álvarez, M.A.; Mayfield, M. City-Scale Residential Energy Consumption Prediction with a Multimodal Approach. *Sci. Rep.* **2025**, *15*, 5313. [CrossRef]
34. Jiang, Q.; Huang, C.; Wu, Z.; Yao, J.; Wang, J.; Liu, X.; Qiao, R. Predicting Building Energy Consumption in Urban Neighborhoods Using Machine Learning Algorithms. *FURP* **2024**, *2*, 6. [CrossRef]
35. Wang, R.; Lu, S.; Li, Q. Multi-Criteria Comprehensive Study on Predictive Algorithm of Hourly Heating Energy Consumption for Residential Buildings. *Sustain. Cities Soc.* **2019**, *49*, 101623. [CrossRef]
36. Liu, K.; Xu, X.; Zhang, R.; Kong, L.; Wang, W.; Deng, W. Impact of Urban Form on Building Energy Consumption and Solar Energy Potential: A Case Study of Residential Blocks in Jianhu, China. *Energy Build.* **2023**, *280*, 112727. [CrossRef]
37. Liu, K.; Xu, X.; Zhang, R.; Kong, L.; Wang, X.; Lin, D. An Integrated Framework Utilizing Machine Learning to Accelerate the Optimization of Energy Efficient Urban Block Forms. *Build. Simul.* **2024**, *17*, 2017–2042. [CrossRef]
38. Wang, W.; Liu, K.; Zhang, M.; Shen, Y.; Jing, R.; Xu, X. From Simulation to Data-Driven Approach: A Framework of Integrating Urban Morphology to Low-Energy Urban Design. *Renew. Energy* **2021**, *179*, 2016–2035. [CrossRef]
39. Feng, Y.; Duan, Q.; Chen, X.; Yakkali, S.S.; Wang, J. Space Cooling Energy Usage Prediction Based on Utility Data for Residential Buildings Using Machine Learning Methods. *Appl. Energy* **2021**, *291*, 116814. [CrossRef]
40. Chen, P.; Wu, Y.; Zhong, H.; Long, Y.; Meng, J. Exploring Household Emission Patterns and Driving Factors in Japan Using Machine Learning Methods. *Appl. Energy* **2022**, *307*, 118251. [CrossRef]
41. Wei, Z.; Zhang, T.; Yue, B.; Ding, Y.; Xiao, R.; Wang, R.; Zhai, X. Prediction of Residential District Heating Load Based on Machine Learning: A Case Study. *Energy* **2021**, *231*, 120950. [CrossRef]
42. Qavidelfardi, Z.; Tahsildoost, M.; Zomorodian, Z.S. Using an Ensemble Learning Framework to Predict Residential Energy Consumption in the Hot and Humid Climate of Iran. *Energy Rep.* **2022**, *8*, 12327–12347. [CrossRef]
43. Luo, C.; Feng, C.; Zhong, H.; Liu, Y.; Dou, M. Design Optimization of Climate-Responsive Rural Residences in Solar Rich Areas Considering Sustainability and Occupant Comfort. *Energy Build.* **2025**, *336*, 115546. [CrossRef]
44. Ye, Z.; Cheng, K.; Hsu, S.-C.; Wei, H.-H.; Cheung, C.M. Identifying Critical Building-Oriented Features in City-Block-Level Building Energy Consumption: A Data-Driven Machine Learning Approach. *Appl. Energy* **2021**, *301*, 117453. [CrossRef]
45. Liu, Y.; Zhang, W.; Liu, W.; Tan, Z.; Hu, S.; Ao, Z.; Li, J.; Xing, H. Exploring the Seasonal Effects of Urban Morphology on Land Surface Temperature in Urban Functional Zones. *Sustain. Cities Soc.* **2024**, *103*, 105268. [CrossRef]

46. Zou, B.; Fan, C.; Li, J.; Wang, M.; Liao, Y.; Zhou, X. Assessing the Impact of Land Use Changes on Urban Heat Risk under Different Development Scenarios: A Case Study of Guangzhou in China. *Sustain. Cities Soc.* **2025**, *130*, 106532. [CrossRef]
47. Zhou, X.; Deng, S.; Cui, Y.; Fan, C. Developing a Co-Benefits Evaluation Model to Optimize Greening Coverage Designs on University Campuses in Hot and Humid Areas. *Energy Build.* **2025**, *328*, 115214. [CrossRef]
48. Cui, Y.; Fan, C.; Zhou, X.; Liao, Y. Analysis of Green Belt Designs in Mitigating Anthropogenic Heat Emission from Buildings by Considering Cooling Efficiency and Nurture Costs. *Indoor Built Environ.* **2024**, *34*, 438–459. [CrossRef]
49. Zou, B.; Fan, C.; Li, J. Quantifying the Influence of Different Block Types on the Urban Heat Risk in High-Density Cities. *Buildings* **2024**, *14*, 2131. [CrossRef]
50. *JGJ 75-2003*; Design Standard for Energy Efficiency of Residential Buildings in Hot Summer and Warm Winter Zone. Ministry of Housing and Urban-Rural Development of China (MOHURD) China Architecture & Building Press: Beijing, China, 2003.
51. Zhang, X.; Chen, Z.; Yue, Y.; Qi, X.; Zhang, C.H. Fusion of Remote Sensing and Internet Data to Calculate Urban Floor Area Ratio. *Sustainability* **2019**, *11*, 3382. [CrossRef]
52. Song, S.; Leng, H.; Xu, H.; Guo, R.; Zhao, Y. Impact of Urban Morphology and Climate on Heating Energy Consumption of Buildings in Severe Cold Regions. *IJERPH* **2020**, *17*, 8354. [CrossRef]
53. Perera, A.T.D.; Javanroodi, K.; Nik, V.M. Climate Resilient Interconnected Infrastructure: Co-Optimization of Energy Systems and Urban Morphology. *Appl. Energy* **2021**, *285*, 116430. [CrossRef]
54. Ratti, C.; Baker, N.; Steemers, K. Energy Consumption and Urban Texture. *Energy Build.* **2005**, *37*, 762–776. [CrossRef]
55. Guangdong Construction Yearbook Compilation Committee. *Guangdong Construction Yearbook 2015*; Guangdong People's Publishing House: Guangzhou, China, 2015; ISBN 978-7-218-10509-3.
56. Wang, Z.; Zhou, R.; Rui, J.; Yu, Y. Revealing the Impact of Urban Spatial Morphology on Land Surface Temperature in Plain and Plateau Cities Using Explainable Machine Learning. *Sustain. Cities Soc.* **2025**, *118*, 106046. [CrossRef]
57. Li, Z.; Ma, J.; Jiang, F.; Zhang, S.; Tan, Y. Assessing the Impacts of Urban Morphological Factors on Urban Building Energy Modeling Based on Spatial Proximity Analysis and Explainable Machine Learning. *J. Build. Eng.* **2024**, *85*, 108675. [CrossRef]
58. Wang, J.; Ren, C. Analysis of residential energy consumption ratio from Guangzhou Asian Games City project. *Water Wastewater Eng.* **2011**, *37*, 92–96. [CrossRef]
59. Demir Dilsiz, A.; Ng, K.; Kämpf, J.; Nagy, Z. Ranking Parameters in Urban Energy Models for Various Building Forms and Climates Using Sensitivity Analysis. *Build. Simul.* **2023**, *16*, 1587–1600. [CrossRef]
60. Fan, C.; Zou, B.; Li, J.; Wang, M.; Liao, Y.; Zhou, X. Exploring the Relationship between Air Temperature and Urban Morphology Factors Using Machine Learning under Local Climate Zones. *Case Stud. Therm. Eng.* **2024**, *55*, 104151. [CrossRef]
61. Chen, Y.; Fan, C.; Nie, Y.; Wu, H.; Zhu, Y.; Liao, Y.; Li, H.; Wu, L.; Lao, M. Integrating Bottom-up Energy Model with Urban Parameterization for Fine-Scale Anthropogenic Heat Estimation. *Case Stud. Therm. Eng.* **2025**, *73*, 106627. [CrossRef]
62. Oh, S.; Ahn, H.; Bae, M.; Kang, J. Development and Analysis of Easy-to-Implement Green Retrofit Technologies for Windows to Reduce Heating Energy Use in Older Residential Buildings. *Sustainability* **2025**, *17*, 3307. [CrossRef]

buildings

MDPI

Article

A* Algorithm for On-Site Collaborative Path Planning in Building Construction Robots

Yuan Fang [1], Jialiang He [1], Xi Wang [1], Wensheng Xu [2], Jung In Kim [3] and Xingbin Chen [4,*]

1 School of Civil and Transportation Engineering, Guangdong University of Technology, No. 100 West Waihuan Road, Panyu District, Guangzhou 510006, China; carolynfang@gdut.edu.cn (Y.F.); 2112209159@mail2.gdut.edu.cn (J.H.); 2112309001@mail2.gdut.edu.cn (X.W.)

2 Zhuhai Huafa Urban Operation and Investment Holdings Co., Ltd., Room 401, No. 9 Lian'an Road, Gongbei District, Zhuhai 519000, China; xuwensheng@hfy12.wecom.work

3 Department of Civil and Environmental Engineering, Kookmin University, 77 Jeongneung-ro, Seongbuk-gu, Seoul 02707, Republic of Korea; jikim07@kookmin.ac.kr

4 College of Water Conservancy and Civil Engineering, South China Agricultural University, No. 483 Wushan Road, Tianhe District, Guangzhou 510642, China

* Correspondence: xbchen@scau.edu.cn

Abstract

This study explores the use of construction robots with collaborative path planning and coordination in complex building construction tasks. Current construction processes involving robots are often fragmented due to their single-task focus, with limited research focused on employing multiple construction robots to collaboratively perform tasks. To address such a challenge, this research proposes an improved A* algorithm for global path planning and obstacle avoidance, combined with the development of a BIM-based grid map of the construction site. The leader–follower method is utilized to guide the robot group in maintaining an optimal formation, ensuring smooth collaboration during construction. The methodology includes formalizing building construction site environments into BIM-based grid maps, path planning, and obstacle avoidance, which allows robot groups to autonomously navigate and complete specific tasks such as concrete, masonry, and decoration construction. The results of this study show that the proposed approach achieves significant reductions in pathlength and operational time of approximately 9% and 10%, respectively, while maintaining safety and efficiency compared with traditional manual methods. This research demonstrates the potential of collaborative construction robot groups to enhance productivity, reduce labor costs, and provide a scalable solution for the intelligent transformation of the construction industry; extends the classical A* algorithm by incorporating obstacle density into the heuristic function; and proposes a new node simplification strategy, contributing to the literature on robot motion planning in semi-structured environments.

Keywords: construction robots; path planning; A* algorithm; robot group formation; smart construction

1. Introduction

The construction industry, as a pillar of the national economy, serves as a critical platform for developing "new quality productivity." However, the construction sectors worldwide have encountered developmental challenges, necessitating transformation and upgrades [1]. As shown in Figures 1 and 2, both the labor force population and the number of construction workers in Mainland China have shown a fluctuating but overall declining trend from 2019 to 2023. Meanwhile, the average age of construction workers has exhibited a clear aging trajectory. In recent years, advancements in intelligent technologies have provided fresh momentum for global manufacturing, with the requirements for intelligence gradually permeating various fields. Smart construction, which integrates automation and informatization, has become a key driver for revolutionizing the construction industry [2,3]. Currently, construction industries across the globe still heavily rely on labor-intensive production and fragmented management methods. These practices result in low production efficiency, significant environmental burdens, and excessive resource consumption [4]. The introduction of the concept of smart construction presents a novel developmental opportunity and trend for the industry. To promote intelligent innovation within the construction sector, incorporating robotics technology into industry transformations and upgrades is imperative. By embedding robotics with intelligence and automation into construction processes, this initiative is poised to exert profound impacts on construction methods, business philosophies, industrial ecosystems, and enterprise development models [5].

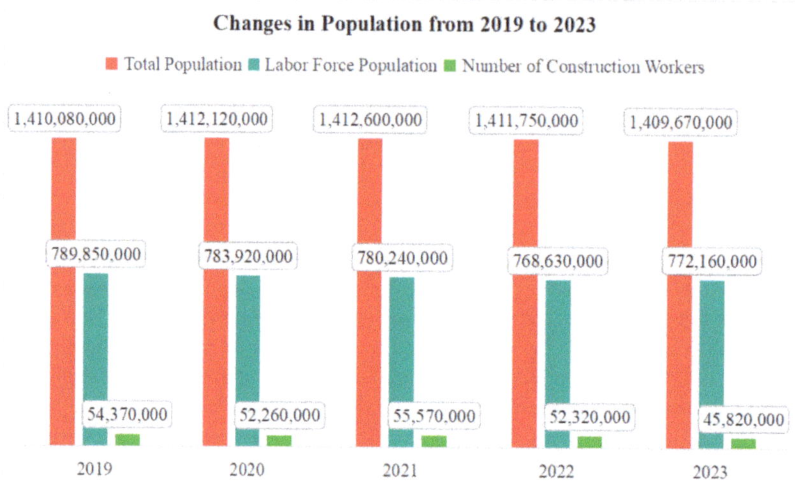

Figure 1. Changes in population from 2019 to 2023.

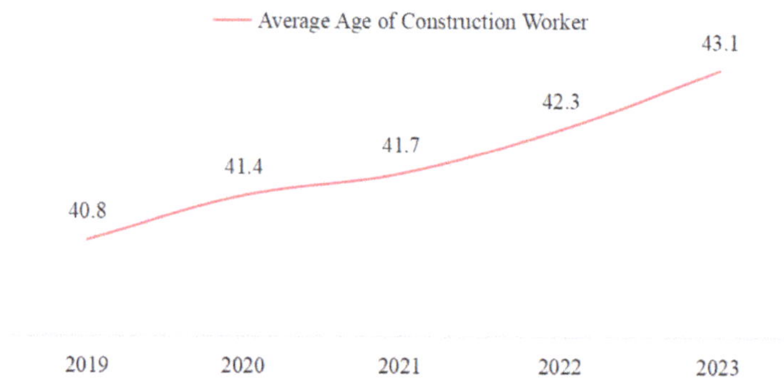

Figure 2. Changes in average age of construction workers from 2019 to 2023.

The number of construction robots used for independent or assistive operations in building projects is gradually increasing in the construction industry. For instance, the MULE bricklaying robot developed in the United States enhances efficiency by assisting workers with bricklaying tasks through a lifting-and-gripping mechanism and a precision positioning system, achieving twice the efficiency of manual labor [6]. Similarly, Australia's Hadrian X robot, equipped with an extendable robotic arm, can autonomously construct the walls of a single building, boasting a work efficiency three times higher than that of humans [7]. However, most building construction robots currently in use are limited to handling single tasks, which continues the fragmented nature of construction processes [8]. Consequently, these robots do not significantly improve efficiency or reduce costs of construction projects. To enhance the level of intelligence in the construction industry, some studies have proposed employing multiple construction robots to collaboratively perform tasks in laboratory settings. This approach aims to address the inefficiencies caused by sequential operations of individual robots but needs to be further utilized, considering its applicability to the real-world built work environment [9,10].

Collaborative operations among construction robots contribute to more organized construction workflows, reducing labor costs and scheduled time, thereby lowering overall project costs. Therefore, this study proposes a leader–follower collaborative operation method, which addresses the path planning challenge for a group of construction robots working collaboratively. The method involves mapping the real-world built work environment of construction robots, using an improved A* algorithm to plan the overall work path for the leader robot, and employing a leader–follower algorithm to coordinate the formation of follower robots. Although numerous studies have improved the A* algorithm and applied it to robotic path planning, few have explored the integration of path planning with coordinated multi-robot operations specifically in the context of construction tasks. This approach enables the construction robot group to perform building construction tasks in an orderly and efficient manner, offering a solution for the application of construction robots in real-world construction scenarios, with the contributions listed as follows:

- The traditional A* algorithm was improved by incorporating obstacle density into the heuristic function and introducing a path-smoothing post-processing step to generate more efficient and realistic trajectories;
- A collaborative path planning framework for construction robot groups was developed based on the leader–follower and BIM-based grid mapping approaches, enabling coordinated task execution in complex construction scenarios;
- The effectiveness (i.e., planning time, planning length, path variation range, and success rate) was validated through real-world construction case studies, which also demonstrated its practical applicability.

2. Literature Review

The advancement of construction robotics heavily relies on effective path planning and robot group formation control. Path planning techniques, such as A*, Dijkstra's algorithm, and Ant Colony Optimization, are essential for ensuring optimal routes in complex environments. Additionally, various group control strategies like the leader–follower, virtual structure, and behavior-based methods are employed to enable collaborative operations in construction tasks. These foundational approaches are integral to enhancing efficiency and safety in construction robot applications.

2.1. Path Planning Algorithms

The safe and efficient operation of construction robots on site relies on selecting a reasonable path to the construction location. Well-planned paths not only enhance construction efficiency and quality but also ensure safety and resource optimization. This is particularly crucial when coordinating multiple construction robots, where collaborative path optimization becomes even more important. Path planning methods can be categorized into global and local path planning, with global path planning being more commonly employed for mobile robots [11,12]. Prominent global path planning algorithms include the node-based A* algorithm, Dijkstra's algorithm, and the Ant Colony Optimization algorithm.

The A* algorithm is a heuristic search method that evaluates the distance relationship between the current search position and the target location in real time, thereby improving search efficiency [13]. Numerous scholars have since made modifications to this algorithm to address various challenges. For instance, Li et al. [14] proposed a hybrid A* algorithm based on BINN, designing an improved bulldozer equipped with monitoring and autonomous path planning capabilities. This framework enables complete path planning, making the algorithm applicable in engineering practices. Similarly, Auh et al. [15] used the A* algorithm to plan the operational paths for cargo-handling robots in cluttered logistics environments, reducing collisions that could damage packaging while minimizing transport workload, time, and distance. Zhang et al. [16] improved the traditional A* algorithm by expanding the number of search nodes and increasing the search directions to 12, effectively addressing issues such as excessive path turns, uneven paths, and limited search ranges. These enhancements ensured optimal paths, improved safety by avoiding collisions, and increased precision, reducing path length. Zheng et al. [17] integrated an improved A* algorithm with the Timed Elastic Band (TEB) algorithm for motion control, enabling construction robots to navigate static and dynamic obstacles in indoor environments.

The Dijkstra algorithm is a classic solution for finding the shortest path in weighted graphs, often used for routing problems or as a subroutine in more complex graph algorithms [18]. Liu et al. [19] applied the Dijkstra algorithm for global indoor path planning of mobile robots and combined it with the Dynamic Window Approach for local path planning, enabling navigation and obstacle avoidance in building indoor environments. Alshammrei et al. [20] designed an optimized, collision-free algorithm based on the Dijkstra algorithm, achieving efficient path planning in obstacle-filled environments. This approach significantly reduced working time and costs for mobile robots.

The Ant Colony Optimization algorithm, proposed by Marco Dorigo, excels in global optimization but suffers from high computational requirements and a tendency to get trapped in local optima. Yen and Cheng [21] improved the traditional Ant Colony Optimization algorithm by integrating it with fuzzy control, enabling better adaptability in dynamic environments. Li et al. [22] optimized the pheromone update and transmission mechanisms in the Ant Colony Optimization algorithm, accelerating convergence and reducing the likelihood of local optima. Wu et al. [23] addressed the slow convergence and low efficiency of traditional Ant Colony Optimization algorithms by introducing a novel heuristic mechanism with directional information, improving the convergence speed, reducing path turns, and enhancing its practicality and efficiency in path planning.

Table 1 provides a comparative analysis of the three mainstream path planning algorithms—A*, Dijkstra, and Ant Colony algorithms—across several key performance indicators, including computational efficiency, scalability in ability to generate paths in complex site environments, and suitability in its practicality under the specific constraints of construction sites. While all three algorithms can successfully generate paths from the starting node to the destination, their characteristics differ significantly. The Ant Colony Optimization algorithm suffers from high computational overhead and a tendency to get stuck in local optima, making it unsuitable for complex construction scenarios. The Dijkstra algorithm ensures accurate shortest-path computation but lacks efficiency due to exhaustive node expansion. In contrast, the A* algorithm balances speed and optimality by leveraging heuristic functions and global information, enabling it to efficiently handle environments with static obstacles. Furthermore, the A* algorithm demonstrates scalability in multi-robot contexts and integrates easily with BIM-based grid maps, making it particularly well-suited for construction robot path planning. Therefore, this study builds upon the improved A* algorithm as the core of the proposed collaborative planning framework.

Table 1. Algorithm performance comparison.

Type of Algorithm	Scalability in Path Generation	Suitability in Path Optimization	Computational Efficiency
A* algorithm	Yes	Yes	Fastest
Dijkstra algorithm	Yes	Yes	Relatively Fast
Ant Colony algorithm	Yes	No	Slow

2.2. Robot Group Formation Control

Robot groups can achieve collaborative tasks, defense, and other desired outcomes through formation control, information sensing, data analysis, task allocation, trajectory planning, and formation stability control technologies. For construction robot groups performing collaborative tasks on construction sites, robot group formation methods can be referenced. Currently, widely applied methods for group formation include the leader–follower, virtual structure, and behavior-based formation methods.

The leader–follower method is one of the most commonly used methods for multi-robot collaborative formation control and is widely applied in robot and drone control fields [24]. The Robotics Automation Exploration and Perception Laboratory at the University of Pennsylvania has laid a solid theoretical foundation for the application of the leader–follower method in multi-robot systems, proposing two formation strategies: the L-ϕ method, which controls both the relative distance and the yaw angle of the robots, and the L-L control method, which only controls the relative distance [25]. In the application of the leader–follower method, Wu et al. [26] proposed using the leader–follower strategy for multi-robot obstacle avoidance formation control. By pre-setting the leader's motion trajectory and adjusting the angle between the leader and followers during their movement, the robots can avoid obstacles, effectively solving the obstacle avoidance problem in multi-robot motion. Wang et al. [27] applied the leader–follower model for formation control of non-holonomic mobile robot groups in environments with unknown obstacles. They designed a bounded obstacle function to form constraints on the distance and orientation during the follower's pursuit of the leader, enabling obstacle avoidance in unknown environments. Park and Yoo [28] proposed a leader–follower method that guarantees the connectivity between mobile robots, employing an obstacle avoidance strategy that maintains the connection between the leader and follower while avoiding obstacles, ensuring global connectivity even in the presence of obstacles. Li et al. [29] developed a novel virtual leader–leader–follower model based on the leader–follower method. In this model, the leader follows a reference trajectory determined virtual leader, and the followers track the desired position indicated by the virtual leader to solve the formation switching problem for mobile robots in environments with unknown obstacles.

The virtual structure method, proposed by Lewis et al. [30] from the University of California, is widely applied in formation flight control for aircraft and satellites [31]. In the virtual structure method, the entire formation is treated as a virtual rigid structure, where each agent is a fixed node in the relative position of the rigid body, and the agents follow the motion of the virtual structure. The group of agents moves with the rigid body in space, causing changes in their positions, but their relative positions remain unchanged, preserving the inherent geometric shape. The virtual structure method transforms the multi-agent formation problem into a tracking problem, offering high control accuracy and feedback on the formation, effectively controlling the overall behavior of the formation group [32]. Cai et al. [33] proposed a distributed model predictive control scheme based on virtual target guidance for controlling unmanned aerial vehicles (UAVs). In this method, each UAV shares information only with its neighbors, reducing the computational burden of UAV formation control and effectively achieving trajectory tracking and obstacle avoidance. Zhen et al. [34] proposed a method for multi-underwater vehicle formation control that combines artificial potential fields with the virtual structure. This combined control method ensures that the underwater vehicles avoid collisions during descent and can form and alter their formation once the desired depth is reached, achieving better mission detection outcomes.

The behavior-based formation method, proposed by Balch and Arkin [35], is a control approach inspired by biomimicry. It essentially involves a collection of behavioral actions such as obstacle avoidance, aggregation, formation change, and formation maintenance. Each action corresponds to a specific control method, with weights assigned to each action. The basic idea is to first design and coordinate the behaviors and then adjust and sum the weights of the behaviors to achieve the desired action set. In practical applications, Cheung et al. [36] proposed a behavior-based architecture to mitigate the impact of external disturbances on follower robots. They introduced a behavior manager that can create multiple behaviors to form a convoy controller combination, enabling the robot formation to maintain its movement even when subjected to disruptive attacks, thereby improving the robustness of the mobile robot team. Chen et al. [37] introduced a behavior-based decision-making path planning method, achieving behavior-based formation control and optimizing the diving paths of underwater robots to save energy and improve task efficiency. Table 2 compares the advantages and disadvantages of the above methods for collaborative formation control.

Table 2. Comparison of formation methods.

Type of Formation Methods	Advantages	Disadvantages
Leader–Follower Method	Simple formation control structure, easy to implement	Highly dependent on the leader
Virtual Structure Method	Clear formation feedback, easy to determine and maintain formation behavior	Restricted in flexibility and adaptability, especially in case of obstacle avoidance
Behavior-Based Formation Method	Strong adaptability, better handle collision avoidance issues	Limited in clearly defining the overall behavior of the formation system, leading to lower stability

As shown in Table 2, the virtual structure method views the formation as a rigid structure, providing clear formation feedback and making it easy to maintain formation. However, since the formation must constantly maintain the same rigid structure, it lacks flexibility and adaptability, and there are certain limitations in the process of obstacle avoidance. This leads to the relatively limited application of the virtual structure method in multi-robot formation control. The advantage of the behavior-based formation method is that it allows for easier implementation of distributed control, and the system is more adaptable, effectively addressing collision avoidance issues. Additionally, the formation can achieve feedback through the perception between robots. However, the integration of behaviors in this method is complex, and the basic local behaviors for a robot to achieve a specified formation are difficult to design, which makes ensuring the stability of formation control hard. Therefore, this method is also rarely used in engineering applications. The leader–follower method, being the most commonly used formation control method, although highly dependent on the leader, has a simple control structure and is easy to implement. In this formation, the leader's desired path is set, and the followers simply follow the leader with a predetermined positional offset to maintain the formation. Among the three mainstream formation control strategies, each presents distinct engineering trade-offs:

- ➢ The leader–follower approach is easy to implement and highly effective in environments with clear navigation paths but is susceptible to leader failure and communication delays, making robustness a key concern.
- ➢ The virtual structure method ensures rigid formation and accurate coordination but suffers from high computational overhead and limited scalability in complex or crowded construction sites.
- ➢ Behavior-based control offers better fault tolerance and flexibility in dynamic environments but may lack formation stability and requires fine-tuned behavior rules.

Considering the semi-structured nature of most construction sites, the leader–follower model offers a practical compromise between control simplicity, system scalability, and formation robustness.

2.3. *Summary of Current Research*

As for path planning algorithms of construction robots, they can currently be divided into global and local path planning algorithms. For mobile robots applied in construction scenarios, global path planning algorithms such as A* algorithm, Dijkstra algorithm, and Ant Colony algorithm are generally used. Among these algorithms, the A* algorithm is particularly suitable for path planning in complex environments due to its faster and more efficient search process. It can also ensure the optimality of the path, even when there are static obstacles available.

In terms of collaborative formation methods for robots, the widely used methods include the leader–follower method, virtual structure method, and behavior-based formation method. However, since the latter two methods have limitations in engineering applications, such as flexibility and stability limitations, this study adopts the leader–follower method, which is easier to implement in construction scenarios.

Despite extensive applications of classical A* algorithm and its variants in robotics, few studies in the construction domain address the unique challenges of multi-robot coordination in cluttered and fragmented environments. To bridge this gap, we adopt a leader–follower formation strategy integrated with an improved A* algorithm that accounts for obstacle density and path smoothness. This approach directly responds to the need for scalable and efficient path planning in real-world construction scenarios by developing a collaborative path planning method for construction robots. The leader robot in the construction robot group uses an improved A* algorithm to plan a smooth path that avoids obstacles, while the follower robots in the group uses the leader–follower method to follow the path. This approach aims to validate the feasibility and applicability of construction robots working on building construction sites.

3. Methodology

This section outlines the methodology employed for analyzing and optimizing the collaborative work of construction robots. First, a detailed examination of collaborative building work scenarios for different types of robots is conducted, highlighting how these robots interact within specific construction environments. The process continues with the formalization of a work map, which is essential for enabling construction robots to navigate and perform tasks within a defined workspace. Finally, the proposed method integrates path planning and obstacle avoidance techniques, with an improved A* algorithm to enhance the robots' ability to efficiently navigate complex environments. The methodology also incorporates a leader–follower approach for coordinating multiple robots in a collaborative setting, ensuring smooth operation and formation maintenance during construction tasks.

3.1. Analysis of Collaborative Work Scenarios for Construction Robots

To use specific construction robots for particular tasks, it is essential not only to select the appropriate type of robot but also to consider the work scenarios in which they will operate. Currently, various types of building construction robots are available on the market, including intelligent follow-up concrete placing robots for concrete construction, floor leveling robots, and floor finishing robots [38]; mortar spraying robots for secondary structure construction [39], block transport robots [40], and bricklaying robots [6]; as well as decoration robots such as putty coating robots, wallpaper laying robots, and wall tile laying robots [41]. While many construction robots have already been commercially deployed, the problem of collaborative work among construction robots requires specific robot combinations to form a group that can work together to complete building construction tasks. The work scenarios for these robots primarily include the following:

Collaborative Work Scenarios for Two Construction Robots: The collaboration between two construction robots is primarily applied to standard floor concrete construction and indoor wall and floor tile installation. For standard floor concrete construction, a floor leveling robot and a floor finishing robot typically work together in collaboration to complete the task. For indoor wall and floor tile installation, a wall tile laying robot and a floor tile laying robot collaborate with a block transport robot to perform the construction tasks.

Collaborative Work Scenarios for Three Construction Robots: Collaboration among three construction robots is typically applied to tasks that are more complex and involve multiple construction processes. Current scenarios where three construction robots work together include standard floor masonry construction and interior decorative coating spraying. Standard floor masonry construction involves collaboration between a block transport robot, a mortar spraying robot, and a bricklaying robot. The block transport robot handles material transportation, while the mortar spraying and bricklaying robots perform the masonry work. Interior decorative coating spraying involves collaboration between putty decoration and finishing robots, and an indoor spraying robot. The putty decoration and finishing robots first complete the basic puttying tasks, after which the indoor spraying robot applies the corresponding paint to finish the interior wall decoration.

3.2. Formalization of Robots' Work Map

To enable construction robots to operate in a specific building working environment, the environment needs to be converted into a work map for the robots, allowing them to recognize and navigate it. In the field of robotics, there are three main types of map representations: metric, topological, and semantic maps [42]. For mobile construction robots, the most commonly used map type is metric maps, specifically the grid type. Due to its model simplicity and the uniqueness of its positional representation, grid maps are frequently used in path planning. A grid map divides the entire environment into several equally sized grids, and different representations are assigned to each grid based on the presence of obstacles. The advantage of using grid maps for path planning is that it allows the map to be drawn with sufficiently fine grids, enabling the representation of contours of arbitrary shapes [43]. Path planning based on grid maps can be performed in three directions—horizontal, vertical, and diagonal—allowing the robot to follow the planned path precisely. The steps to map the building construction site environment and build a grid map are shown in Figure 3:

➢ *Step 1:* Collect building blueprint information to determine the geometry and location of each floor of the building.

- *Step 2:* Use Building Information Modeling (BIM) technology to draw the building's geometric and locational information diagram and create the construction robot's working environment model.
- *Step 3:* Set the grid plane parameters and mapping rules based on the construction robot's working environment model.
- *Step 4:* Complete the actual mapping of the work map based on the grid map and mapping rules, thus obtaining the construction robot's working grid map.

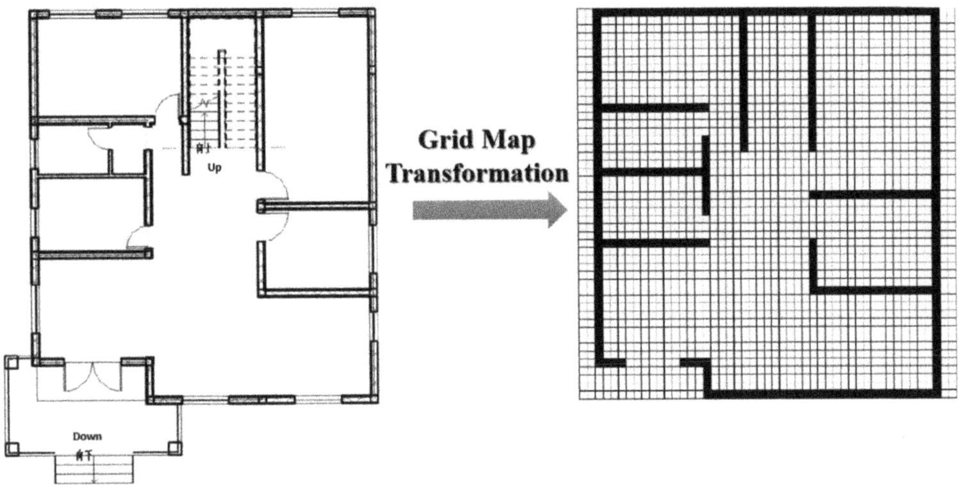

Figure 3. Formalization of building's grid map.

In step 3, the mapping rules are primarily as follows: The white areas in the grid map are defined as free grids, indicating obstacle-free regions, while the areas with obstacles are represented by black grids, which indicate regions where the robot cannot pass. To facilitate understanding by computers, this environmental information is represented using a 0–1 matrix. A 0 represents free space, and a 1 represents an obstacle space.

3.3. Path Planning and Obstacle Avoidance for the Leader Robot Based on A Algorithm*

In building construction site operations, it is essential to ensure that the construction robots select a reasonable and effective path. Based on global path planning, the path must be optimized according to the site's environment, allowing the construction robots to effectively avoid obstacles while selecting the shortest operational path.

Traditional A Algorithm:* for path planning in the construction industry, typically in static environments, the A* algorithm not only provides the shortest path but also ensures high search efficiency. The A* algorithm performs a directional search, utilizing the evaluation function in Formula (1) to calculate the distance from each node to the destination, thereby determining whether the node can be part of the shortest path.

$$f(n) = g(n) + h(n) \tag{1}$$

where $f(n)$ is the estimated distance from the initial node to the target node through node N, $g(n)$ represents the actual distance from the initial node to node N in the state space, and $h(n)$ is the estimated distance from node N to the target node along the optimal path.

For the A* algorithm to guarantee optimality, the key lies in the selection of $h(n)$. If $h(n)$ represents the estimated distance from node N to the target node, then $h*(n)$ is the actual distance from node N to the target node. The strength of the heuristic function is measured by how closely $h(n)$ approximates $h*(n)$. Ideally, $h(n) \leq h*(n)$, and the more heuristic information $h(n)$ contains, the closer it will be to $h*(n)$, resulting in a more optimal path.

Improved A Algorithm for Path Planning and Obstacle Avoidance:* While the traditional A* algorithm performs well in many cases, it also has certain limitations. In building construction sites, where numerous obstacles may exist, using the traditional A* algorithm for path planning may result in a locally optimal solution, making it unable to effectively address obstacle avoidance.

To adapt to the practical requirements of construction site operations, this study incorporates obstacle considerations into the path planning process. The algorithm is designed to favor paths with fewer obstacles during the search process. Dynamic weights are introduced into the evaluation function, factoring in the proportion of obstacles between the starting node and the target node. By incorporating an obstacle rate, the influence of the heuristic weight can be adjusted, encouraging the algorithm to prioritize paths with fewer obstacles. The modified evaluation function, i.e., $f'(n)$, derived based on the above improvement approach, is as follows:

$$f\prime(n) = g(n) + e^{(q \times p)} \times h(n) \quad (2)$$

where e^x is the exponential function in mathematics, and q represents the influence of the obstacle rate on the weight of the heuristic factor. This parameter adjusts the impact of obstacle density on the heuristic function. To guide the algorithm towards regions with fewer obstacles, a larger q value increases the impact of obstacles on path selection. If the path contains more obstacles, the heuristic function $h(n)$ is multiplied by a larger weight, prompting the algorithm to choose paths with fewer obstacles.

The value of q is experimentally adjusted within a given environment. Typically, q is adjusted within a range of 1 to 10, and the optimal value for a specific environment is identified through iterative fine-tuning based on experimental results. To tune the dynamic weight parameter q in the heuristic function, we conducted empirical testing within a reasonable interval. If q is too small (e.g., approaching 0), the obstacle density term becomes insignificant, causing the algorithm to degenerate into the standard A* without improvements. Conversely, an excessively large q (e.g., greater than 10) leads to overly conservative planning, resulting in unnecessarily long detours or increased planning failures in dense environments. Therefore, we adopted a common tuning range of $q \in [1,10]$, which balances global optimality and obstacle avoidance. Through simulation trials, we found that $q = 5$ produced the most effective results in terms of smoothness, path length, and planning success rate. This value was thus used in our implementation.

The parameter p represents the obstacle rate between the current and target node, calculated using Formula (3). In the local grid map between the current and target node, the number of obstacle grids is denoted as O. The current node's coordinates are (x^n, y^n), and the target node's coordinates are (x_g, y_g). When calculating the distance in each direction, one is added to ensure that the path length is not zero and to maintain stability in the calculations.

$$p = \frac{O}{(|x_n - x_g| + 1)(|y_n - y_g| + 1)}, p \in (0, 1) \quad (3)$$

Path Improvement: The traditional A* algorithm often results in paths with excessive turning and redundant nodes. To address this issue, this study proposes a key node extraction method to filter out redundant nodes from the A* algorithm's planned path. This approach shortens the path length, reduces turning nodes, and makes the overall working path smoother, thereby improving the operational efficiency of construction robots. The optimization steps are as follows:

- ➢ Obtain the complete set of nodes $U = \{Pi|1 \leq I \leq n\}$ from the A* algorithm's planned path, where P_1 represents the starting node and P_n represents the end node of the path. Create an initial key node set $V = \{P_1, P_n\}$, which initially includes only the starting and ending nodes of the path and is used to store the optimized key nodes.
- ➢ Starting from P_1, draw a straight line connecting the nodes sequentially in U, i.e., P_2, P_3, . . ., P_i, and check if the straight line P_1P_i intersects any obstacles:
- ➢ If the line P_1P_i does intersect an obstacle, the node P_{i-1} is deemed a critical node and a necessary turning node in the path and is added to the set V.
- ➢ If the line P_1P_i does not intersect any obstacles, nodes P_2, . . ., P_{i-1} are considered redundant. *Continue* extending the line from P_i to subsequent nodes in U until it reaches the end node P_n, adding all identified key nodes to V.
- ➢ After selecting the key nodes, the set V contains all critical nodes. Sequentially connect all nodes in V to globally optimize the path, i.e., remove redundant nodes and smooth the planned path.

3.4. Path Planning for Collaborative Construction Robot Groups

The leader robot uses the A* algorithm for path planning and obstacle avoidance. On this basis, the leader–follower approach is introduced to control the collaborative formation of the construction robot group. The follower robots follow the leader robot and actively avoid obstacles. This approach enables the robot group to follow the leader robot along a defined path while maintaining a specific formation, ensuring orderliness during the construction process. The leader robot employs the *Improved A* Algorithm* for global path planning, generating a reference path for the construction robot group. Based on the real-time positioning of the leader robot and the desired formation, the relative distance and angle of each follower robot to the leader are used to calculate the position of the virtual leader robot. The position of the virtual leader is then transmitted to the corresponding follower robot as a target node, guiding each follower robot to move toward its respective target.

The model of the leader robot in the global coordinate system is shown in Figure 4, where

- ➢ The global coordinate system is represented by X–O–Y.
- ➢ The local coordinate system of the leader robot is represented by X_L–O_L–Y_L.
- ➢ The position of the leader robot in the global coordinate system is defined as (x_L, y_L, θ_L), where
- ➢ x_L, y_L are the coordinates of the leader robot in the global coordinate system,
- ➢ θ_L is the angle between the orientation of the leader robot and the X-axis in the global coordinate system.
- ➢ The velocity vector of the leader robot in the global coordinate system is expressed as [44].

$$\begin{bmatrix} \dot{x} \\ \dot{y} \\ \dot{\theta} \end{bmatrix} = \begin{bmatrix} cos\theta_L & -sin\theta_L & 0 \\ sin\theta_L & cos\theta_L & 0 \\ 0 & 0 & 1 \end{bmatrix} \begin{bmatrix} v_x \\ v_y \\ \omega \end{bmatrix} \tag{4}$$

where

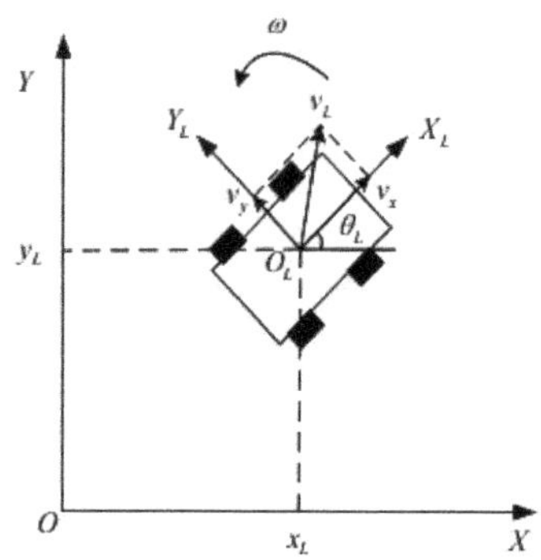

Figure 4. Mathematical model of the leader robot.

- ➢ $\dot{x}$, $\dot{y}$, and $\dot{\theta}$ represent the leader robot's linear velocities in the x and y directions and angular velocity in the global coordinate system;
- ➢ v_x, v_y, and ω represent the leader robot's linear velocities in the x and y directions and angular velocity in the local coordinate system.

As shown in Figure 5, assume the position of the follower robot is (x_F, y_F, θ_F). Based on the formation requirements, let the position of the virtual leader be (x_V, y_V, θ_V). The expected distance between the virtual leader and the leader robot is L_{L-F}, and the expected angle is φ_{L-F}. Using the expected distance and angle, the position of the virtual leader in the global coordinate system can be expressed as

$$\begin{cases} x_V = x_L + L_{L-F}\cos(\varphi_{L-F} + \theta_L) \\ y_V = y_L + L_{L-F}\sin(\varphi_{L-F} + \theta_L) \\ \theta_V = \theta_L \end{cases} \quad (5)$$

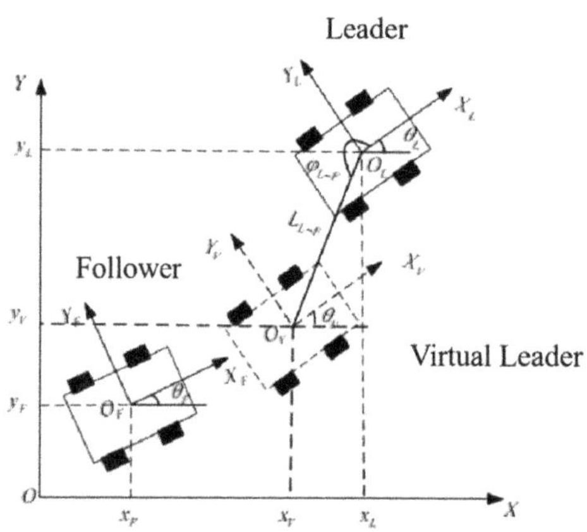

Figure 5. Mathematical model of the follower robot.

The proposed objective is to ensure that the position of the follower converges to that of the virtual leader robot. In other words, the follower's final destination should align with the virtual leader robot's position, rather than the position of the leader robot. Therefore, the position error function is constructed as follows:

$$\begin{cases} x_d = x_V - x_F \\ y_d = y_V - y_F \\ \theta_d = \theta_V - \theta_F \end{cases} \quad (6)$$

where x_d, y_d, and θ_d represent the position error between the virtual leader and the follower robot in the global coordinate system. Using a transformation matrix, the position error in the global coordinate system can be converted into the local coordinate system of the follower robot as follows:

$$\begin{bmatrix} d_x \\ d_y \\ d_\theta \end{bmatrix} = \begin{bmatrix} cos\theta_F & sin\theta_F & 0 \\ -sin\theta_F & cos\theta_F & 0 \\ 0 & 0 & 1 \end{bmatrix} \begin{bmatrix} x_d \\ y_d \\ \theta_d \end{bmatrix} \tag{7}$$

where d_x, d_y, and d_θ represent the position error between the virtual leader and the follower robot in the local coordinate system of the follower robot. Substituting Formula (5) into Formula (6) and simplifying, the error function in the follower robot's local coordinate system can be expressed as

$$\begin{bmatrix} d_x \\ d_y \\ d_\theta \end{bmatrix} = \begin{bmatrix} cos\theta_F(x_V - x_F) + sin\theta_F(y_V - y_F) \\ sin\theta_F(x_V - x_F) + cos\theta_F(y_V - y_F) \\ \theta_V - \theta_F \end{bmatrix} \tag{8}$$

Substituting the position formula of the virtual leader robot in the global coordinate system into the error function in the local coordinate system, i.e., Formula (5) into Formula (8), and differentiating Formula (8) yields

$$\begin{cases} \dot{d_x} = v_L cosd_\theta + \omega_F d_y - v_F - \omega_L L_{L-F} \sin(\varphi_{L-F} + d_\theta) \\ \dot{d_y} = v_L - \omega_F d_x + \omega_L L_{L-F} \cos(\varphi_{L-F} + d_\theta) \\ \dot{d_\theta} = \omega_V - \omega_F \end{cases} \tag{9}$$

As can be seen from Formula (9), the robot formation control problem is transformed into a trajectory tracking problem for the follower robot relative to the virtual leader robot. When the error is minimized, the follower robot converges to the position of the virtual leader. At this point, the follower robot moves to the virtual leader's position using the optimal linear and angular velocities computed, thus completing the formation of the robot group.

4. Case Studies

This study uses masonry work on the standard floor of a building construction project as the tested collaborative work scenario for construction robots, a widely adopted and mature application scenario in the domain. The masonry construction scenario was selected as the case study due to its representative characteristics. First, the task involves repetitive block transport, mortar spraying, and bricklaying, which facilitates the measurement and control of robotic behaviors. Second, masonry sites typically include a cluttered and semi-structured environment, with common construction obstacles such as stacked bricks, scaffolding, and temporary barriers. Lastly, masonry tasks inherently require multi-robot collaboration, as the supply of materials must be coordinated with the bricklaying operation. These features make masonry an ideal and challenging scenario to evaluate the effectiveness of coordinated robot path planning in real-world conditions.

The selected robots include a block transport robot, a mortar spraying robot, and a bricklaying robot to perform path planning for the masonry task collaboratively. The masonry work that the construction robots need to complete is highlighted within the red frame in the BIM model shown in Figure 6, where the work environment contains various obstacles of different sizes.

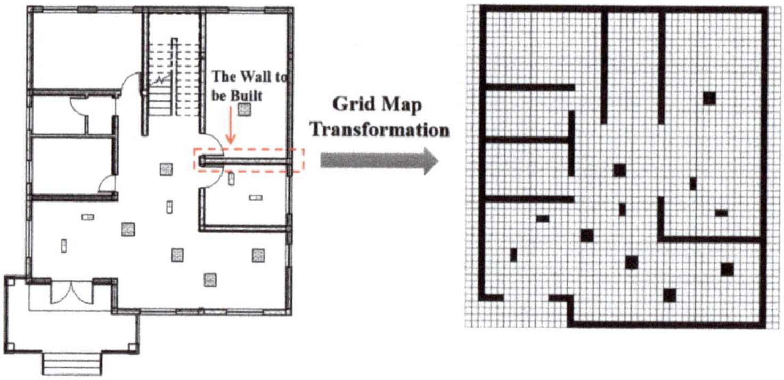

Figure 6. BIM model and grid conversion environment for the construction robot.

Figure 6 shows the grid transformation of the BIM model, where the environmental information is represented by black-and-white grid cells. The blank grids are referred to as free grids, indicating areas without obstacles where the robot can pass through, while the black grids represent obstacle grids, indicating areas with obstacles that the construction robot cannot pass through. Figure 7 converts the environmental information into a computer-readable language represented by a 0–1 matrix. In this matrix, 1 represents an obstacle-occupied space, which the robot cannot pass, while 0 represents a free space, where the robot can move freely.

	1	2	3	4	5	6	7	8	9	10	11	12	13	14	15	16	17	18	19	20	21	22	23	24	25	26	27	28	29	30	31	32	33	34	35	36	37	38	39	40	41	42	43	44	45	46	47	48	49	50
1	0	0	0	0	0	0	0	0	0	0	0	0	0	0	0	0	1	1	1	1	1	1	1	1	1	1	1	1	1	1	1	1	1	1	1	1	1	1	1	1	1	1	1	1	1	1	1	1	0	0
2	0	0	0	0	0	0	0	0	0	0	0	0	0	0	0	0	1	0	0	0	0	0	0	0	0	0	0	0	0	0	0	0	0	0	0	0	0	0	0	0	0	0	0	0	0	0	0	1	0	0
3	0	0	0	0	0	0	0	0	0	0	0	0	0	0	0	0	1	0	0	0	0	0	0	0	0	0	0	0	0	0	0	0	0	0	0	0	0	0	0	0	0	0	0	0	0	0	0	1	0	0
4	0	0	0	0	0	0	0	0	0	0	0	0	0	0	0	0	1	0	0	0	0	0	0	0	0	0	0	0	0	0	0	0	0	0	0	0	0	0	0	0	0	0	0	0	0	0	0	1	0	0
5	0	0	1	1	1	1	0	0	0	0	0	0	0	1	1	1	1	0	0	0	0	0	0	0	0	0	0	0	0	0	0	0	1	1	0	0	0	0	0	0	0	0	0	0	0	0	0	1	0	0
6	0	0	1	0	0	0	0	0	0	0	0	0	0	0	0	0	0	0	0	0	0	0	0	0	0	0	0	0	0	0	0	0	1	1	0	0	0	0	0	0	0	0	0	0	0	0	0	1	0	0
7	0	0	1	0	0	0	0	0	0	0	0	0	0	0	0	0	0	0	0	0	0	0	0	0	0	0	0	0	0	0	0	0	0	0	0	0	0	0	0	0	0	0	0	0	0	0	0	1	0	0
8	0	0	1	0	0	0	0	0	0	0	0	0	0	0	0	0	0	0	0	0	0	0	0	0	0	0	0	0	0	0	0	0	0	0	0	0	0	0	0	0	0	0	0	0	0	0	0	1	0	0
9	0	0	1	0	0	0	0	0	0	0	0	0	0	0	0	0	0	0	0	0	0	0	0	0	0	0	0	0	0	0	0	0	0	0	0	0	0	0	0	0	0	0	0	0	0	0	0	1	0	0
10	0	0	1	0	0	0	0	0	0	0	0	0	0	0	0	0	0	0	0	0	0	0	0	0	0	0	0	0	0	0	0	0	0	0	0	0	0	0	0	0	0	1	1	0	0	0	0	1	0	0
11	0	0	1	0	0	0	0	0	0	0	0	0	0	0	0	0	0	0	0	0	0	0	0	0	0	0	1	1	0	0	0	0	0	0	0	0	0	0	0	0	0	1	1	0	0	0	0	1	0	0
12	0	0	1	0	0	0	0	1	0	0	0	0	0	0	0	0	0	0	0	0	0	0	0	0	0	0	1	1	0	0	0	0	0	0	0	0	0	0	0	0	0	0	0	0	0	0	0	1	0	0
13	0	0	1	0	0	0	0	1	0	0	0	0	0	0	0	0	0	0	0	0	0	0	0	0	0	0	0	0	0	0	0	0	0	0	0	0	0	0	0	0	0	0	0	0	0	0	0	1	0	0
14	0	0	1	0	0	0	0	0	0	0	0	0	0	0	0	0	0	0	0	0	0	0	0	0	0	0	0	0	0	0	0	0	0	0	0	0	0	0	0	0	0	0	0	0	0	0	0	1	0	0
15	0	0	1	0	0	0	0	0	0	0	0	0	0	0	0	0	0	0	1	1	0	0	0	0	0	0	0	0	0	0	1	1	1	1	1	1	1	1	1	1	1	1	1	1	1	1	1	1	0	0
16	0	0	1	0	0	0	0	0	0	0	0	0	0	0	0	0	0	0	1	1	0	0	0	0	0	0	0	0	0	0	1	0	0	0	0	0	0	0	0	0	0	0	0	0	0	0	0	1	0	0
17	0	0	1	0	0	0	0	0	0	0	0	0	0	0	0	0	0	0	0	0	0	0	0	0	0	0	0	0	0	0	1	0	0	0	0	0	0	0	0	0	0	0	0	0	0	0	0	1	0	0
18	0	0	1	0	0	0	0	0	0	0	0	1	1	0	0	0	0	0	0	0	0	0	0	0	0	0	0	0	0	0	1	0	0	0	0	0	0	0	0	0	0	0	0	0	0	0	0	1	0	0
19	0	0	1	0	0	0	0	0	0	0	0	0	0	0	0	0	0	0	0	0	0	0	0	0	0	1	0	0	0	0	1	0	0	0	0	0	0	0	0	0	1	1	0	0	0	0	0	1	0	0
20	0	0	1	0	0	0	0	0	0	0	0	0	0	0	0	0	0	0	0	0	0	0	0	0	0	1	0	0	0	0	1	0	0	0	0	0	0	0	0	0	0	0	0	0	0	0	0	1	0	0
21	0	0	1	1	1	1	1	1	1	1	1	1	1	1	1	1	1	0	0	0	0	0	0	0	0	0	0	0	0	0	1	0	0	0	0	0	0	0	0	0	0	0	0	0	0	0	0	1	0	0
22	0	0	1	0	0	0	0	0	0	0	0	0	0	0	0	0	0	0	0	0	0	0	0	0	0	0	0	0	0	0	0	0	0	0	0	0	0	0	0	0	0	0	0	0	0	0	0	1	0	0
23	0	0	1	0	0	0	0	0	0	0	0	0	0	0	0	0	0	0	0	0	0	0	0	0	0	0	0	0	0	0	0	0	0	0	0	0	1	0	0	0	0	0	0	0	0	0	0	1	0	0
24	0	0	1	0	0	0	0	0	0	0	0	0	0	0	0	0	0	0	0	0	0	0	0	0	0	0	0	0	0	0	0	0	0	0	0	0	1	0	0	0	0	0	0	0	0	0	0	1	0	0
25	0	0	1	0	0	0	0	0	0	0	0	0	0	0	0	0	1	0	0	0	0	0	0	0	1	1	0	0	0	0	0	0	0	0	0	0	0	0	0	0	0	0	0	0	0	0	0	1	0	0
26	0	0	1	0	0	0	0	0	0	0	0	0	0	0	0	0	1	0	0	0	0	0	0	0	1	1	0	0	0	0	0	0	0	0	0	0	0	0	0	0	0	0	0	0	0	0	0	1	0	0
27	0	0	1	0	0	0	0	0	0	0	0	0	0	0	0	0	1	0	0	0	0	0	0	0	0	0	0	0	0	0	0	0	0	0	0	0	0	0	0	0	0	0	0	0	0	0	0	1	0	0
28	0	0	1	0	0	0	0	0	0	0	0	0	0	0	0	0	1	0	0	0	0	0	0	0	0	0	0	0	0	0	0	0	0	0	0	0	0	0	0	0	0	0	0	0	0	0	0	1	0	0
29	0	0	1	1	1	1	1	1	1	1	1	1	1	1	1	1	1	0	0	0	0	0	0	0	0	0	0	0	0	0	0	0	0	0	0	0	0	0	0	0	0	0	0	0	0	0	0	1	0	0
30	0	0	1	0	0	0	0	0	0	0	0	0	0	0	0	0	1	0	0	0	0	0	0	0	0	0	0	0	0	0	0	0	0	0	0	0	0	0	0	0	0	0	0	0	0	0	0	1	0	0
31	0	0	1	0	0	0	0	0	0	0	0	0	0	0	0	0	1	0	0	0	0	0	0	0	0	0	0	0	0	0	0	0	0	0	0	0	0	0	0	0	0	0	0	0	0	0	0	1	0	0
32	0	0	1	0	0	0	0	0	0	0	0	0	0	0	0	0	1	0	0	0	1	0	0	0	0	0	0	0	0	0	1	0	0	0	0	0	0	0	0	0	0	0	0	0	0	0	0	1	0	0
33	0	0	1	0	0	0	0	0	0	0	0	0	0	0	0	0	0	0	0	0	1	0	0	0	0	0	0	0	0	0	1	0	0	0	0	0	0	0	0	0	0	0	0	0	0	0	0	1	0	0
34	0	0	1	0	0	0	0	0	0	0	0	0	0	0	0	0	0	0	0	0	1	0	0	0	0	0	0	0	0	0	1	0	0	0	0	0	0	0	0	0	0	0	0	0	0	0	0	1	0	0
35	0	0	1	0	0	0	0	0	0	0	0	0	0	0	0	0	0	0	0	0	1	0	0	0	0	0	0	0	0	0	1	0	0	0	0	0	0	0	0	0	0	0	0	0	0	0	0	1	0	0
36	0	0	1	0	0	0	0	0	0	0	0	0	0	0	0	0	0	0	0	0	1	0	0	0	0	0	0	0	0	0	1	0	0	0	0	0	0	0	0	0	0	0	0	0	0	0	0	1	0	0
37	0	0	1	0	0	0	0	0	0	0	0	0	0	0	0	0	0	0	0	0	1	0	0	0	0	0	0	0	0	0	1	0	0	0	0	0	0	0	1	1	0	0	0	0	0	0	0	1	0	0
38	0	0	1	1	1	1	1	1	1	1	1	1	1	1	1	1	1	0	0	0	1	0	0	0	0	0	0	0	0	0	1	0	0	0	0	0	0	0	1	1	0	0	0	0	0	0	0	1	0	0
39	0	0	1	0	0	0	0	0	0	0	0	0	0	0	0	0	0	0	0	0	1	0	0	0	0	0	0	0	0	0	1	0	0	0	0	0	0	0	0	0	0	0	0	0	0	0	0	1	0	0
40	0	0	1	0	0	0	0	0	0	0	0	0	0	0	0	0	0	0	0	0	1	0	0	0	0	0	0	0	0	0	1	0	0	0	0	0	0	0	0	0	0	0	0	0	0	0	0	1	0	0
41	0	0	1	0	0	0	0	0	0	0	0	0	0	0	0	0	0	0	0	0	1	0	0	0	0	0	0	0	0	0	1	0	0	0	0	0	0	0	0	0	0	0	0	0	0	0	0	1	0	0
42	0	0	1	0	0	0	0	0	0	0	0	0	0	0	0	0	0	0	0	0	1	0	0	0	0	0	0	0	0	0	1	0	0	0	0	0	0	0	0	0	0	0	0	0	0	0	0	1	0	0
43	0	0	1	0	0	0	0	0	0	0	0	0	0	0	0	0	0	0	0	0	1	0	0	0	0	0	0	0	0	0	1	0	0	0	0	0	0	0	0	0	0	0	0	0	0	0	0	1	0	0
44	0	0	1	0	0	0	0	0	0	0	0	0	0	0	0	0	0	0	0	0	1	0	0	0	0	0	0	0	0	0	1	0	0	0	0	0	0	0	0	0	0	0	0	0	0	0	0	1	0	0
45	0	0	1	0	0	0	0	0	0	0	0	0	0	0	0	0	0	0	0	0	1	0	0	0	0	0	0	0	0	0	1	0	0	0	0	0	0	0	0	0	0	0	0	0	0	0	0	1	0	0
46	0	0	1	0	0	0	0	0	0	0	0	0	0	0	0	0	0	0	0	0	1	0	0	0	0	0	0	0	0	0	1	0	0	0	0	0	0	0	0	0	0	0	0	0	0	0	0	1	0	0
47	0	0	1	0	0	0	0	0	0	0	0	0	0	0	0	0	0	0	0	0	1	0	0	0	0	0	0	0	0	0	1	0	0	0	0	0	0	0	0	0	0	0	0	0	0	0	0	1	0	0
48	0	0	1	0	0	0	0	0	0	0	0	0	0	0	0	0	0	0	0	0	1	0	0	0	0	0	0	0	0	0	1	0	0	0	0	0	0	0	0	0	0	0	0	0	0	0	0	1	0	0
49	0	0	1	0	0	0	0	0	0	0	0	0	0	0	0	0	0	0	0	0	1	0	0	0	0	0	0	0	0	0	1	0	0	0	0	0	0	0	0	0	0	0	0	0	0	0	0	1	0	0
50	0	0	1	1	1	1	1	1	1	1	1	1	1	1	1	1	1	1	1	1	1	1	1	1	1	1	1	1	1	1	1	1	1	1	1	1	1	1	1	1	1	1	1	1	1	1	1	1	0	0

Figure 7. Grid map conversion to 0–1 matrix.

4.1. A* Algorithm-Based Path Planning and Obstacle Avoidance Model for the Leader Robot

An improved A* algorithm is employed for path planning to enable the leader robot to find an optimal work path. The planning process is presented in Figure 8. The starting node A is set at coordinates (11, 3), and the end node B is set at coordinates (45, 26), corresponding to the location of the wall construction task in as discussed in Section 4.2. The improved A* algorithm is implemented for path planning simulation using MATLAB (Version R2022b).

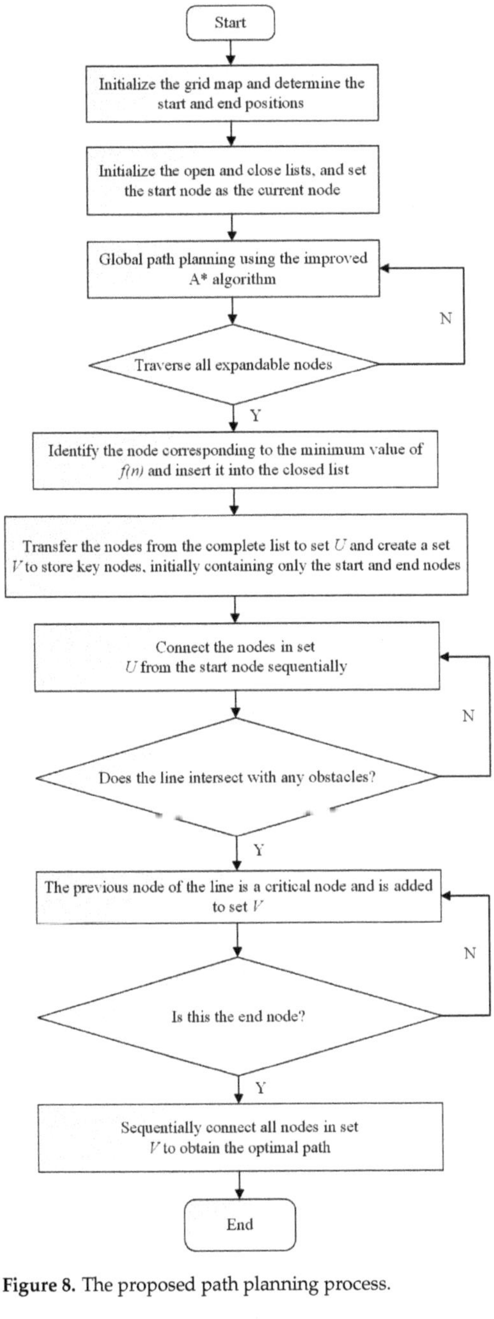

Figure 8. The proposed path planning process.

This study improves the A* algorithm. The leader robot uses the improved A* algorithm for global path planning based on the path planning process, resulting in an optimal global path that avoids obstacles to the greatest extent. The evaluation function of the improved A* algorithm used in the path planning experiment is as follows:

$$f(n) = g(n) + e^{(5\times p)} \times h(n) \tag{10}$$

The obstacle rate p is determined jointly by the obstacle environment in Figure 9 and Formula (3). To match the real construction environment, the influence value q of the obstacle rate on the heuristic factor weight is adjusted within the range of 1 to 10 in the environment provided in Figure 9. Given the presence of a certain number of obstacles in the environment and to better balance the path length and safety in the improved A* algorithm, a middle value of q (i.e., q = 5) is chosen, which results in an ideal effect.

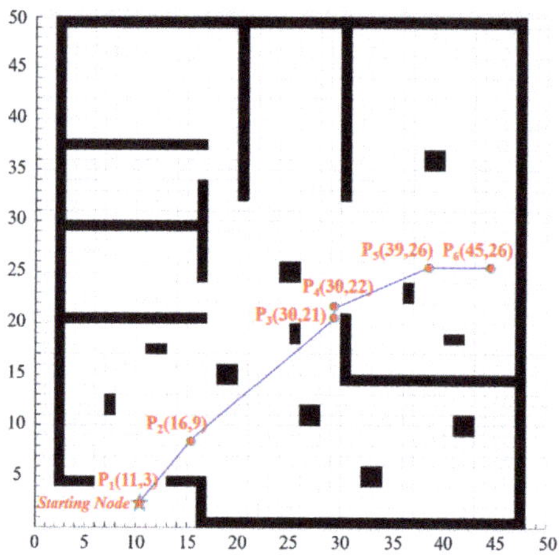

Figure 9. The key nodes and improved path.

After the algorithm finds the path, to simplify and smoothen the path, key nodes are identified and redundant nodes are removed, thus obtaining the optimal path. The set V is used to store the key nodes in the path, and the coordinates of the key nodes of the optimized path are presented in Table 3.

Table 3. Key node coordinates.

Key Node	Horizontal Coordinate Value	Vertical Coordinate Value
P_1 (Star Node)	11	3
P_2	16	9
P_3	30	21
P_4	30	22
P_5	39	26
P_6 (End Node)	45	26

The nodes in key node set V are indicated in red in Figure 9. After constructing the key node set, the key nodes are sequentially connected from the starting node. The resulting smooth curve is the planned optimal path, which is a work path for the leader robot that avoids obstacles and is smoother. As shown in Figure 9, the construction robot starts from the doorway and, during while approaching the work area, effectively avoids the obstacle-dense areas in the working environment. The planned path using the improved

A* algorithm avoids obstacles and maintains a smooth path, making the work process safer and more efficient.

4.2. Collaborative Path Planning for a Group of Construction Robots

In this case, the mortar spraying robot acts as the leader robot, while the bricklaying and block transport robots function as Followers 1 and 2, respectively. After the leader robot uses the improved A* algorithm for path planning, the followers, under the leader–follower control strategy, follow the leader while maintaining a specific formation to perform collaborative tasks. The starting node of the leader robot is A with coordinates (11, 3), and the end node B is at coordinates (45, 26). The initial heading angle of the leader robot is set as $\theta_L = \pi/4$, so the leader robot's position is (11, 3, $\pi/4$). The distance from the leader robot to the followers is set as $L_{(L\text{-}F)} = 2$, with the initial angles $\varphi_{(L\text{-}F)}$ set to $\pi/2$ and $3\pi/4$, respectively. Therefore, the initial positions of follower robots C and D are (9, 3, $\pi/4$) and (11, 1, $\pi/4$), respectively. The relevant parameters for the simulation are set as listed in Table 4. The simulation process is shown in Figure 10.

Table 4. Experimental parameters.

Parameter Types	Value
Initial heading angle of the leader robot θ_L	$\pi/4$
Initial heading angle of the follower robot θ_F	$\pi/4$
Desired distance between the leader and the followers L_{L-F}	2 m
Initial angle between the leader and follower φ_{L-F}	$\pi/2$
Initial angle between the leader and follower φ_{L-F}	$3\pi/4$
Maximum linear velocity of the robot	0.7 m/s
Minimum linear velocity of the robot	0 m/s
Maximum linear acceleration of the robot	0.4 m/s^2
Minimum linear acceleration of the robot	0 m/s^2

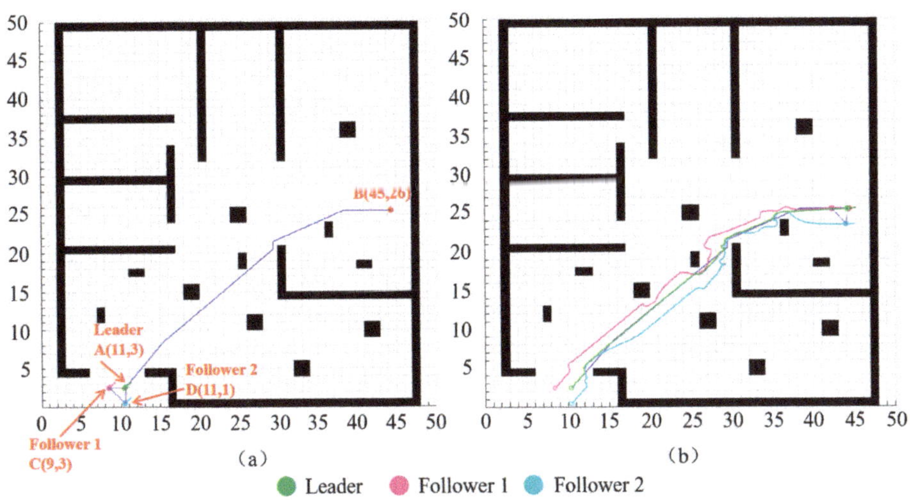

Figure 10. Path planning for collaborative operation of robot group based on leader–follower approach. (**a**) The starting nodes of the leader, Follower 1 and 2 robots (**b**) The paths generated for the leader, Follower 1 and 2 robots.

During operation, to maintain the formation, the follower robots are guided by a virtual leader. The virtual leader maintains a constant distance $L_{L-F} = 2$ from the leader robot while guiding its movement. The follower robots utilize the error optimization function, as shown in Formula (5), to minimize the error, allowing the follower robots to converge at the position of the virtual leader. At this point, the follower robots calculate the optimal linear and angular velocities to move to the position of the virtual leader, as shown in Figure 11, thus completing the collaborative formation path planning task for the robot group.

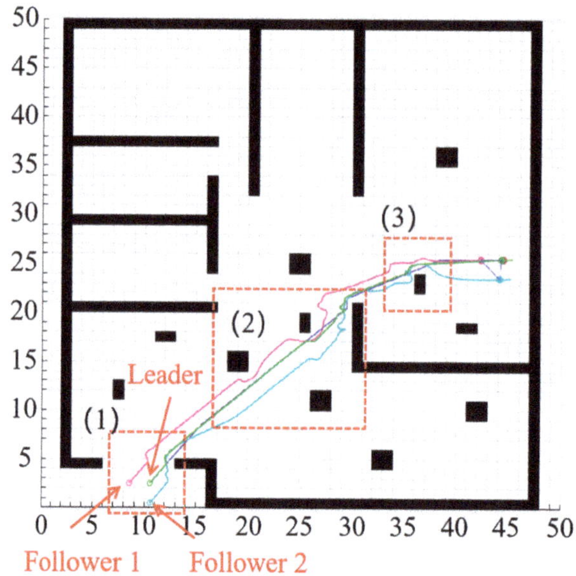

Figure 11. The main obstacle areas of the robot group working path.

4.3. Discussion

The changes in the heading angles of the leader and follower robots are shown in Figure 10. The three construction robots start from the initial position with a heading angle $\theta = \pi/4$. During movement, *Follower 2* first reaches the obstacle area *(1)*. To avoid the wall on the right side of the obstacle area *(1)*, *Follower 2* adjusts its angle to achieve obstacle avoidance. Simultaneously, to maintain the formation of the robot group, *Follower 1* also adjusts its angle accordingly. From the start node to the end node, due to the presence of multiple obstacles in the environment, the follower robots (Followers 1 and 2) continuously adjust their heading angles by a certain magnitude to ensure that the leader robot operates along the optimal path while maintaining formation. After passing the obstacle area *(1)*, the heading angle changes for both the leader and follower robots, gradually stabilizes, and becomes smoother.

As the leader robot moves along the planned optimal path, which avoids obstacle-dense areas and is relatively smooth, the overall change in its heading angle is small, with a significant heading angle adjustment only occurring near the obstacle area *(2)*. In contrast, the two follower robots, while maintaining a fixed formation and avoiding obstacles, experience more frequent heading angle adjustments throughout the movement. However, as the number of obstacles decreases or after passing the obstacle area *(3)* and nearing the end node, the variations in heading angle of the follower robots gradually decrease, eventually stabilizing and converging to a fixed value as they reach the end node and complete the motion. This indicates that even when encountering obstacles, the robot group can maintain a consistent working angle throughout the collaborative operation, ultimately reaching the target position efficiently and effectively.

The variation in the distance between the robot group and the leader robot during movement is shown in Figures 12 and 13. At the start of the motion, the leader and follower robots maintain the desired distance $L_{(L-F)} = 2$. During the movement, when the follower robots encounter obstacles, their distances from the leader robot fluctuate to varying degrees. In obstacle-dense areas, the fluctuation in distance between the follower robots and the leader robot is more pronounced. Once the robots move away from the obstacles, the fluctuation in distance gradually stabilizes, eventually returning to the initially set desired distance $L_{(L-F)} = 2$ at the end node. This demonstrates that during collaborative operation, even when obstacles are encountered, the distance between the leader and the follower robots may experience variation. Nevertheless, the robot group can maintain a consistent distance throughout the operation, ultimately reaching the target position effectively.

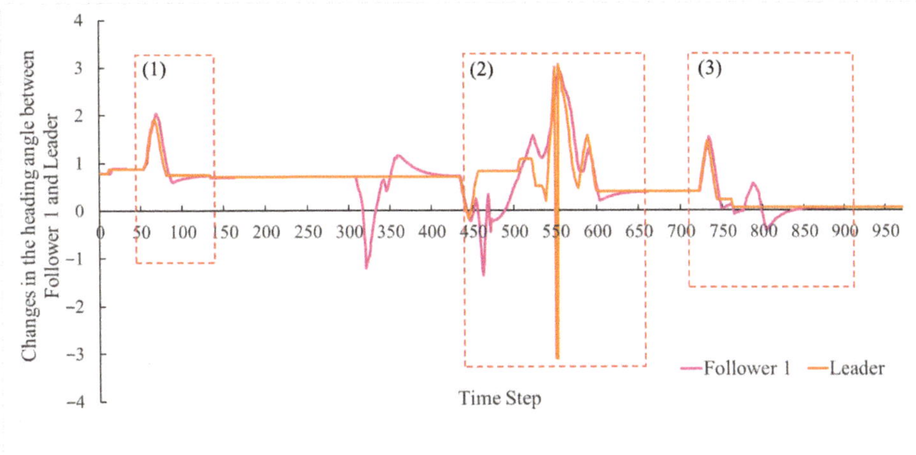

(a)

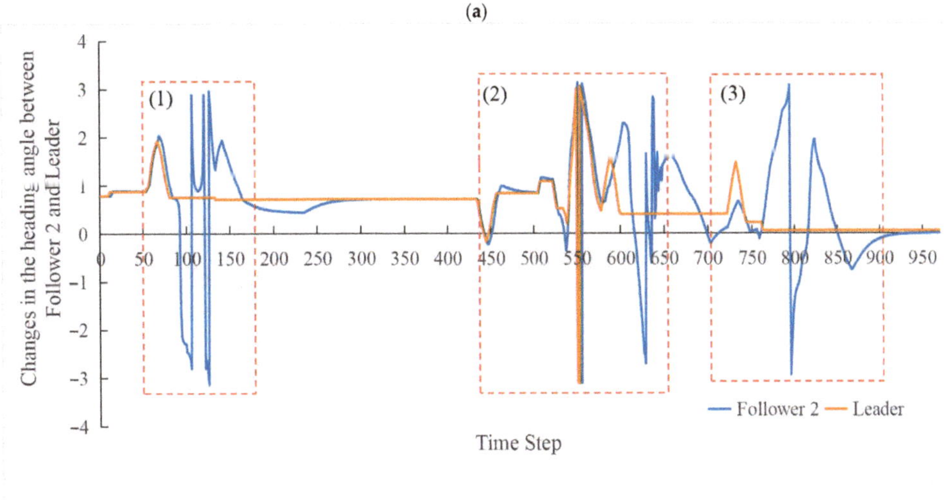

(b)

Figure 12. Changes in the heading angles of the leader and follower robots. (**a**) Changes in the heading angle between Follower 1 and Leader. (**b**) Changes in the heading angle between Follower 2 and Leader.

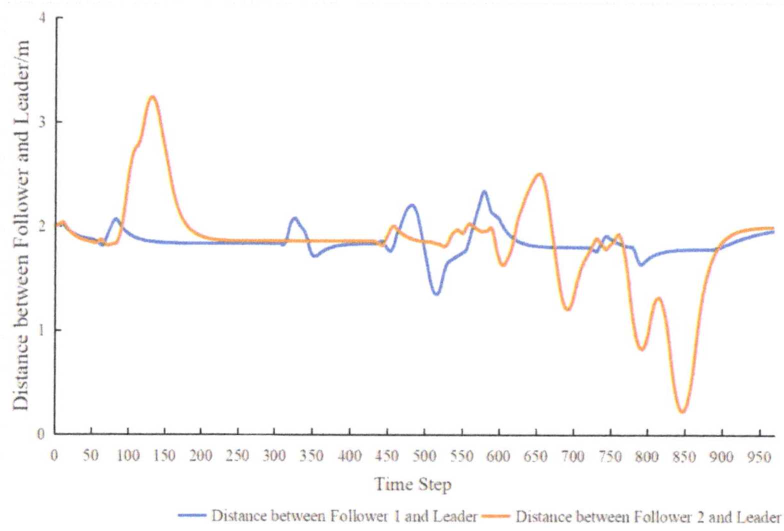

Figure 13. The variation in distance *L* between the leader and follower robots.

In summary, applying this leader–follower algorithm, after setting the initial heading angle and desired distance between the leader robot and the follower robots, the follower robots may temporarily deviate in angle and distance during operation due to obstacle avoidance. However, after passing through obstacles, the follower robots and the leader robot can return to the initially set heading angle and distance to continue moving. Similarly, when the position of the leader robot changes, both follower robots can make appropriate position adjustments to maintain the formation. Ultimately, the robots converge to the designated positions with a specific angle and distance.

While this study focuses primarily on planning efficiency, path smoothness, and successful task completion rate, it does not incorporate additional performance indicators such as energy consumption, communication overhead, or robustness under dynamic obstacles. These dimensions will be explored in future studies to enhance the practical applicability of the proposed system.

This framework provides a feasible pathway for enabling collaborative operations in construction environments, addressing the limitations of traditional single-robot deployment. From a theoretical perspective, the proposed method contributes to the body of knowledge on multi-agent systems by operationalizing leader–follower dynamics in semi-structured, obstacle-rich environments such as construction sites. This study also supports the integration of decentralized coordination and modular formation control into the robotics-in-construction paradigm.

In terms of practical implications, the reduced path variation, shorter planning time, and higher task success rates imply enhanced productivity and operational safety. While quantitative cost data were not collected in this study, the use of automated coordination among robots holds promise for reducing labor dependence and minimizing rework, especially in repetitive masonry tasks. This aligns with current industry trends toward automation under labor shortages.

Despite these contributions, certain limitations must be acknowledged. The method relies on accurate obstacle mapping and assumes static obstacle conditions. In highly dynamic environments, real-time perception and replanning capabilities would be necessary to ensure robustness. Additionally, the leader–follower model, while effective for coordination, may face challenges in scalability and flexibility when applied to larger, more heterogeneous robot teams. Future work will focus on addressing these challenges by incorporating adaptive formation strategies and dynamic reconfiguration mechanisms.

4.4. Verification

To verify that the proposed on-site collaborative path planning for construction robots conformed to those in practical engineering applications, this study invited four experts (with an average of eight years of experience) to evaluate the proposed method. A grid map containing obstacles, obtained through environmental mapping, was provided to them. Under the condition of predefined start and end node locations, the participants were asked to manually plan a path on the provided grid map. The results of the manual path planning were compared with those generated based on the improved A* algorithm proposed in this study. A fitting analysis was conducted to compare the coordinate variations in the manually optimized path and the one generated by the proposed method (Figure 14).

$$R^2 = 1 - \frac{RSS}{TSS} \quad (11)$$

$$RSS = \sum_{1}^{6} (y_i - y_s)^2 \quad (12)$$

$$TSS = \sum_{1}^{6} (y_i - \overline{y})^2 \quad (13)$$

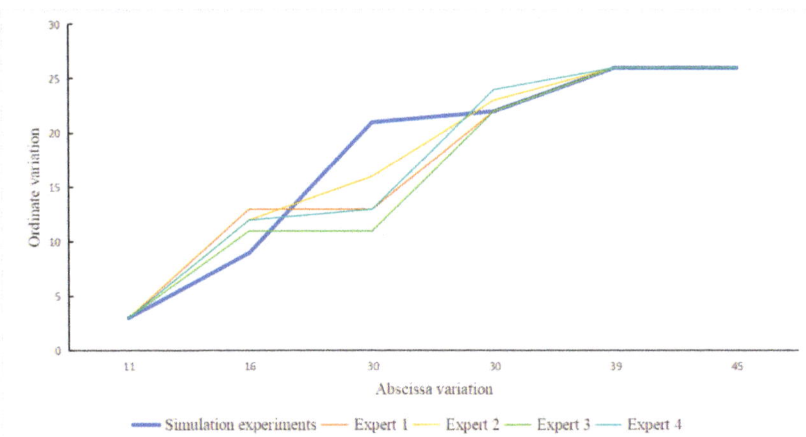

Figure 14. Coordinate variations for the paths generated using the proposed method and manually.

To analyze the degree of fit between the manual results and the one generated by the proposed method, this study uses the coefficient of determination R^2 for evaluation. This metric is widely used to measure the degree of difference between predicted and actual values [45].

Here, y_i represents the i-th vertical coordinate of the key nodes in the path planned by the technicians, and $\overline{y}$ is the mean of the vertical coordinates of the key nodes in the manually planned path. y_s represents the i-th vertical coordinate of the key nodes in the path generated from the proposed method. The R^2 value, calculated using Formula (10), is used to evaluate the fit between the manually planned paths and those generated using the proposed method. The Total Sum of Squares (TSS) represents the sum of the squared differences between the dependent variable and its mean. The Residual Sum of Squares (RSS) represents the sum of the squared differences between the values generated using the proposed method and the dependent variable. The closer the R^2 value is to 1, the better the fit. Conversely, an R^2 value closer to 0 or negative indicates a poor fit.

To simplify calculations and balance precision with efficiency, this study substitutes the coordinates of the selected key nodes from the proposed method (Figure 11) and the corresponding node coordinates obtained manually (Figure 14) into Formulas (11) and (12) for the calculation of R^2. The fitting results are shown in Table 5. The goodness-of-fit values between the paths planned by the four experts based on their experience and the ones from the proposed method are approximately between 0.7 and 0.9. This indicates that the collaborative path planning for construction robots meets the requirements of actual construction site operations, demonstrating a certain degree of conformity.

Table 5. Path goodness-of-fit calculation.

Investigator	TSS	RSS	R^2
Expert 1	415	64	0.84
Expert 2	414	35	0.91
Expert 3	452	104	0.77
Expert 4	444	84	0.81

Table 6 compares the average performance metrics between the path generated using the proposed method and manually, including task completion time, path length, path variation, and task success rate. The results show that collaborative path planning for construction robots demonstrates more significant advantages. The optimal path planned in the proposed method reduced planning time by nearly 10% compared to the manually planned paths. The path generated by the proposed method not only exhibited smaller variations and smoother transitions but also reduced path length by approximately 9%. This demonstrates that the proposed method in this study can also effectively achieve optimal values. Furthermore, a risk of collisions or near collisions exists in the manually planned paths, resulting in a task success rate of approximately 75%, where the path generated by the proposed method achieved a better result.

Table 6. Comparison between the simulated experimental path and the expert planned path.

Path Planning Method	Planning Time (s)	Path Length/m	Path Variation Range	Success Rate
The proposed method	547.7	22	Minor	100%
Manual method	604.9	24.3	Significant	75%

The path generated in the proposed method reduced planning time by nearly 10% compared to the path planned by experts. Additionally, the path generated in the proposed method exhibited smaller variations, greater smoothness, and an approximately 9% shorter path length. These results indicate that the proposed method in this study can achieve a better optimal level. Furthermore, due to the potential for collisions or near-collisions in manual path planning for construction robots, the success rate of completing the assigned task was smaller at approximately 75%. Although the validation involved four experts with an average of eight years of construction experience, this qualitative approach was adopted to ensure that the algorithmic path planning aligns with real-world operational logic. Their evaluations helped confirm that the simulation results are not only computationally optimal but also practically feasible. Future studies will incorporate quantitative validation using multiple benchmarks and real-world robotic deployment scenarios to enhance methodological robustness.

5. Conclusions

Addressing challenges such as limited automation, high dependency on manual labor, and prolonged construction timelines in traditional processes, researchers have increasingly focused on integrating construction robots to perform specific tasks. This approach not only seeks to improve construction efficiency but also aims to enhance automation within the industry, thereby offering innovative solutions for industry transformation. Beyond individual robots, the introduction of robot groups into construction tasks offers a more effective solution for handling complex, labor-intensive operations. This strategy significantly reduces labor dependence, increases efficiency, standardizes processes, and lowers overall construction costs. Additionally, the collaborative nature of robot groups ensures better coordination, leading to safer, more streamlined construction operations.

This study primarily investigates the path planning challenges associated with the collaborative operation of construction robot groups. By introducing an improved A* algorithm for the path planning and obstacle avoidance of leader robots and employing a leader–follower approach to guide the follower robots, this study offers a framework for collaborative navigation. The developed solution enables construction robot groups to autonomously identify and follow the safe and optimal paths within a construction environment. This research makes contributions to the field of smart construction and enhances the understanding of collaborative path planning and formation control for multi-robot systems operating in building construction settings.

Theoretical Contribution: This work extends the classical A* algorithm by incorporating obstacle density into the heuristic function and proposes a new node simplification strategy, contributing to the literature on robot motion planning in semi-structured environments.

Methodological Contribution: The integration of BIM-based grid mapping with multi-robot formation control demonstrates a systematic framework applicable to indoor masonry scenarios, providing a replicable simulation-to-field workflow.

Practical Contribution: Comparative results reveal that the proposed method shortens path length and reduces execution time by approximately 9–10% compared to expert-defined paths and increases the task completion success rate from 75% to 100%, highlighting practical advantages for labor-intensive and repetitive on-site tasks.

While this study demonstrates the feasibility and validity of collaborative path planning in simulated construction scenarios, several real-world constraints have not yet been incorporated into the current framework. These include sensor measurement errors, real-time communication latency between robots, and dynamic hazards such as unexpected obstacle appearances or human interference. To address these challenges, future research should aim to integrate robust perception technologies, such as millimeter-wave radar and computer vision systems, to enhance environmental awareness and obstacle detection in dynamic jobsite conditions. Additionally, incorporating these uncertainties into both simulation and field experiments will enable the development of more resilient and deployable planning frameworks for construction robotics. The application of artificial intelligence in decision-making could further optimize robot collaboration, improving both operational efficiency and autonomous capabilities on active construction sites.

The improved A* algorithm in this study focuses on minimizing global path length as a key indicator of efficiency; we acknowledge that this metric alone does not fully represent real-world task performance. In practical construction settings, the total task completion time is influenced by additional factors such as the robot's acceleration and deceleration profiles, turning angles, and the logical sequence of construction tasks. Future studies should therefore consider a multi-objective optimization framework that incorporates these temporal and kinematic constraints to better reflect real-world operational demands.

This study is involved in the ongoing development of more intelligent, automated construction processes, with potential implications for improving productivity, safety, and sustainability within the construction industry.

Author Contributions: This study was conducted through a collaborative effort among all authors, each contributing to different aspects of the research. The roles and responsibilities of each author are outlined as follows: Y.F.: Writing—Original Draft, Methodology, Investigation, and Conceptualization; J.H.: Writing—Review and Editing, Data Curation, Software, and Visualization; X.W.: Software; W.X.: Methodology, Data Curation, and Validation; J.I.K.: Software and Validation; X.C. (Corresponding Author): Writing—Review and Editing, Funding Acquisition, and Project Administration. All authors have read and agreed to the published version of the manuscript.

Funding: The research described in this paper was conducted with the support of the Guangdong Natural Science Foundation (Grant #2022A1515110888), Guangzhou Natural Science Foundation (Grant #2023A04J1560), Guangdong Postgraduate Education Foundation (Education Development Grants #2024ANLK_024), Guangdong Science and Technology Innovation Program for Water Resources (Grant #2025-02), and South China Agricultural University Foundation (Education Development Grants #K23157/2022). The authors appreciate the support.

Data Availability Statement: The original contributions presented in this study are included in the article. Further inquiries can be directed to the corresponding author.

Conflicts of Interest: Author Wensheng Xu was employed by the company Zhuhai Huafa Urban Operation and Investment Holdings Co., Ltd. The remaining authors declare that the research was conducted in the absence of any commercial or financial relationships that could be construed as a potential conflict of interest.

References

1. Zhu, Z.; Ning, S. Corporate digital transformation and strategic investments of construction industry in China. *Heliyon* **2023**, *9*, e17879. [CrossRef]
2. Guo, Q.; Zhong, J. The effect of urban innovation performance of smart city construction policies: Evaluate by using a multiple period difference-in-differences model. *Technol. Forecast. Soc. Change* **2022**, *184*, 122003. [CrossRef]
3. Lindsay, N.; Buehling, R.K.; Sun, L. A Sequential Task Addition Distributed Assignment Algorithm for Multi-Robot Systems. *J. Intell. Robot. Syst.* **2021**, *102*, 51. [CrossRef]
4. Zhao, X.; Cheah, C.C. BIM-based indoor mobile robot initialization for construction automation using object detection. *Autom. Constr.* **2023**, *146*, 104647. [CrossRef]
5. Gharbia, M.; Chang-Richards, A.; Lu, Y.; Zhong, R.Y.; Li, H. Robotic technologies for on-site building construction: A systematic review. *J. Build. Eng.* **2020**, *32*, 101584. [CrossRef]
6. Melenbrink, N.; Werfel, J.; Menges, A. On-site autonomous construction robots: Towards unsupervised building. *Autom. Constr.* **2020**, *119*, 103312. [CrossRef]
7. Pan, W.; Li, R.; Bock, T. Design and synthesis of the localization system for the onsite construction robot. In *ISARC. Proceedings of the International Symposium on Automation and Robotics in Construction, Online, 27–29 October 2020*; IAARC Publications: Edinburgh, UK, 2020; Volume 37, pp. 1501–1507.
8. Wang, X.; Wang, S.; Menassa, C.C.; Kamat, V.R.; McGee, W. Automatic high-level motion sequencing methods for enabling multi-tasking construction robots. *Autom. Constr.* **2023**, *155*, 105071. [CrossRef]
9. Wang, D.; Deng, H. Multirobot coordination with deep reinforcement learning in complex environments. *Expert Syst. Appl.* **2021**, *180*, 115128. [CrossRef]
10. Ye, X.; Guo, H.; Luo, Z. Two-stage task allocation for multiple construction robots using an improved genetic algorithm. *Autom. Constr.* **2024**, *165*, 105583. [CrossRef]
11. Fang, W.; Liao, Z.; Bai, Y. Improved ACO algorithm fused with improved Q-Learning algorithm for Bessel curve global path planning of search and rescue robots. *Robot. Auton. Syst.* **2024**, *182*, 104822. [CrossRef]
12. Zhang, J.; Zhu, Z.; Xue, Y.; Deng, Z.; Qin, H. Local path planning of under-actuated AUV based on VADWA considering dynamic model. *Ocean Eng.* **2024**, *310*, 118705. [CrossRef]

13. Hart, P.E.; Nilsson, N.J.; Raphael, B. A Formal Basis for the Heuristic Determination of Minimum Cost Paths. *Ieee Trans. Syst. Sci. Cybern.* **1968**, *4*, 100–107. [CrossRef]
14. Li, R.; Zhou, C.; Dou, Q.; Hu, B. Complete coverage path planning and performance factor analysis for autonomous bulldozer. *J. Field Robot.* **2022**, *39*, 1012–1032. [CrossRef]
15. Auh, E.; Kim, J.; Joo, Y.; Park, J.; Lee, G.; Oh, I.; Pico, N.; Moon, H. Unloading sequence planning for autonomous robotic container-unloading system using A-star search algorithm. *Eng. Sci. Technol. Int. J.* **2024**, *50*, 101610. [CrossRef]
16. Zhang, Z.; Wang, S.; Zhou, J. A-star algorithm for expanding the number of search directions in path planning. In Proceedings of the 2021 2nd International Seminar on Artificial Intelligence, Networking and Information Technology (AINIT), Shanghai, China, 15–17 October 2021.
17. Zheng, H.; Dai, M.; Zhang, Z.; Xia, Z.; Zhang, G.; Jia, F. The Navigation Based on Hybrid A Star and TEB Algorithm Implemented in Obstacles Avoidance. In Proceedings of the 2023 29th International Conference on Mechatronics and Machine Vision in Practice (M2VIP), Queenstown, New Zealand, 21–24 November 2023.
18. Dijkstra, E.W. A note on two problems in connexion with graphs. *Numer. Math.* **1959**, *1*, 269–271. [CrossRef]
19. Liu, L.; Lin, J.; Yao, J.; He, D.; Zheng, J.; Huang, J.; Shi, P. Path Planning for Smart Car Based on Dijkstra Algorithm and Dynamic Window Approach. *Wirel. Commun. Mob. Comput.* **2021**, *2021*, 8881684. [CrossRef]
20. Alshammrei, S.; Boubaker, S.; Kolsi, L. Improved Dijkstra Algorithm for Mobile Robot Path Planning and Obstacle Avoidance. *Comput. Mater. Contin.* **2022**, *72*, 5939–5954. [CrossRef]
21. Yen, C.; Cheng, M. A study of fuzzy control with ant colony algorithm used in mobile robot for shortest path planning and obstacle avoidance. *Microsyst. Technol.* **2018**, *24*, 125–135. [CrossRef]
22. Li, J.; Wang, H. Research on AUV Path Planning Based on Improved Ant Colony Algorithm. In Proceedings of the 2020 IEEE International Conference on Mechatronics and Automation (ICMA), Beijing, China, 13–16 October 2020.
23. Wu, L.; Huang, X.; Cui, J.; Liu, C.; Xiao, W. Modified adaptive ant colony optimization algorithm and its application for solving path planning of mobile robot. *Expert Syst. Appl.* **2023**, *215*, 119410. [CrossRef]
24. Wang, P.K.C. Navigation strategies for multiple autonomous mobile robots moving in formation. *J. Robot. Syst.* **1991**, *8*, 177–195. [CrossRef]
25. Desai, J.P.; Ostrowski, J.; Kumar, V. Controlling formations of multiple mobile robots. In Proceedings of the 1998 IEEE International Conference on Robotics and Automation (Cat. No.98CH36146), Leuven, Belgium, 20 May 1998.
26. Wu, X.; Wang, S.; Xing, M. Observer-Based Leader-Following Formation Control for Multi-Robot with Obstacle Avoidance. *IEEE Access* **2019**, *7*, 14791–14798. [CrossRef]
27. Wang, Y.; Wang, D.; Yang, S.; Shan, M. A Practical Leader–Follower Tracking Control Scheme for Multiple Nonholonomic Mobile Robots in Unknown Obstacle Environments. *IEEE Trans. Control Syst. Technol.* **2019**, *27*, 1685–1693. [CrossRef]
28. Park, B.S.; Yoo, S.J. Connectivity-maintaining obstacle avoidance approach for leader-follower formation tracking of uncertain multiple nonholonomic mobile robots. *Expert Syst. Appl.* **2021**, *171*, 114589. [CrossRef]
29. Li, X.; Liu, X.; Wang, G.; Wu, S.; Li, W. Adaptive virtual leader–leader–follower based formation switching for multiple autonomous tracked mobile robots in unknown obstacle environments. *J. Frankl. Inst.* **2024**, *361*, 106762. [CrossRef]
30. Lewis, M.A.; Kar-Han, T. High Precision Formation Control of Mobile Robots Using Virtual Structures. *Auton. Robot.* **1997**, *4*, 387–403. [CrossRef]
31. Beard, R.W.; Lawton, J.; Hadaegh, F.Y. A coordination architecture for spacecraft formation control. *IEEE Trans. Control Syst. Technol.* **2001**, *9*, 777–790. [CrossRef]
32. Ding, W.; Zhang, L.; Zhang, G.; Wang, C.; Chai, Y.; Yang, T.; Mao, Z. Research on obstacle avoidance of multi-AUV cluster formation based on virtual structure and artificial potential field method. *Comput. Electr. Eng.* **2024**, *117*, 109250. [CrossRef]
33. Cai, Z.; Wang, L.; Zhao, J.; Wu, K.; Wang, Y. Virtual target guidance-based distributed model predictive control for formation control of multiple UAVs. *Chin. J. Aeronaut.* **2020**, *33*, 1037–1056. [CrossRef]
34. Zhen, Q.; Wan, L.; Li, Y.; Jiang, D. Formation control of a multi-AUVs system based on virtual structure and artificial potential field on SE(3). *Ocean Eng.* **2022**, *253*, 111148. [CrossRef]
35. Balch, T.; Arkin, R.C. Behavior-based formation control for multirobot teams. *IEEE Trans. Robot. Autom.* **1998**, *14*, 926–939. [CrossRef]
36. Cheung, C.; Rawashdeh, S.; Mohammadi, A. Jam Mitigation for Autonomous Convoys via Behavior-Based Robotics. *Appl. Sci.* **2022**, *12*, 9863. [CrossRef]
37. Chen, G.; Shen, Y.; Qu, N.; He, B. Path planning of AUV during diving process based on behavioral decision-making. *Ocean Eng.* **2021**, *234*, 109073. [CrossRef]
38. Ding, L.; Jiang, W.; Zhou, Y.; Zhou, C.; Liu, S. BIM-based task-level planning for robotic brick assembly through image-based 3D modeling. *Adv. Eng. Inform.* **2020**, *43*, 100993. [CrossRef]
39. Cho, J.; Lee, J.; Kim, Y.; Shin, K. An analysis model for wind resistance performance of automated exterior wall painting robots in apartment buildings. *KSCE J. Civ. Eng.* **2014**, *18*, 909–919. [CrossRef]

40. Jung, J.; Lee, S.; Myung, H. Indoor Mobile Robot Localization and Mapping Based on Ambient Magnetic Fields and Aiding Radio Sources. *IEEE Trans. Instrum. Meas.* **2015**, *64*, 1922–1934. [CrossRef]
41. Yu, S.; Ryu, B.; Lim, S.; Kim, C.; Kang, M.; Han, C. Feasibility verification of brick-laying robot using manipulation trajectory and the laying pattern optimization. *Autom. Constr.* **2009**, *18*, 644–655. [CrossRef]
42. Ben Salah, I.; Kramm, S.; Demonceaux, C.; Vasseur, P. Summarizing large scale 3D mesh for urban navigation. *Robot. Auton. Syst.* **2022**, *152*, 104037. [CrossRef]
43. Thrun, S. Learning metric-topological maps for indoor mobile robot navigation. *Artif. Intell.* **1998**, *99*, 21–71. [CrossRef]
44. Dierks, T.; Jagannathan, S. Control of Nonholonomic Mobile Robot Formations: Backstepping Kinematics into Dynamics. In Proceedings of the 2007 IEEE International Conference on Control Applications, Singapore, 1–3 October 2007.
45. Zhang, H.; Wu, Z. The general goodness-of-fit tests for correlated data. *Comput. Stat. Data Anal.* **2022**, *167*, 107379. [CrossRef]

buildings

MDPI

Article

Optimization of Multi-Parameter Collaborative Operation for Central Air-Conditioning Cold Source System in Super High-Rise Buildings

Jiankun Yang [1,*], Aiqin Xu [2], Lingjun Guan [3] and Dongliang Zhang [4,*]

1 School of Future Transportation, Guangzhou Maritime University, Guangzhou 510725, China
2 Engineering Company Public Office, Nanjing Fiberglass Research & Design Institute Co., Ltd., Nanjing 210012, China; xuaiqin325@163.com
3 Guangzhou Shijie Energy-Saving Technology Co., Ltd., Guangzhou 510440, China; guanlingjun@sjest.com
4 School of Ocean Equipment Engineering, Guangzhou Maritime University, Guangzhou 510725, China
* Correspondence: yangjiankun@gzmtu.edu.cn (J.Y.); zdl_101@163.com (D.Z.)

Abstract

This paper proposes a hybrid integer optimization method based on the Whale Optimization Algorithm (WOA) for the asymmetric central air conditioning chiller system of a 530-m super high-rise building in Guangzhou. Firstly, a three-hidden-layer multilayer perceptron (MLP) chiller model based on 16,276 sets of measured data and a gradient boosting regression cooling tower model based on 21,369 sets of operating condition data were constructed, achieving high-precision modeling of the energy consumption of all equipment in the chiller system. Secondly, a hybrid encoding strategy of "threshold truncation + continuous relaxation" was proposed to integrate discrete on-off states and continuous operating parameters into WOA, and a three-layer constraint repair mechanism was designed to ensure the physical feasibility of the optimization process and the safe operation of equipment. Verification across three load scenarios—low, medium, and high—showed that the optimized system's energy efficiency ratio (EER) increased by 15.01%, 12.61%, and 11.86%, respectively, with energy savings of 12.91%, 11.18%, and 10.58%. The annual rolling optimization results showed that the average EER increased from 5.07 to 5.88 (16.1%), with energy savings ranging from 8.59% to 18.92%. Sensitivity analysis indicated that pump quantity is the most influential parameter affecting system energy consumption, with an additional pump reducing it by 1.1%. The optimization method proposed in this paper meets the minute-level real-time scheduling requirements of building automation systems and provides an implementable solution for energy-saving optimization of central air conditioning chiller systems in super high-rise buildings.

Keywords: central air conditioning chiller system; asymmetric configuration; hybrid integer optimization; Whale Optimization Algorithm; real-time scheduling; energy-saving optimization

1. Introduction

Central air-conditioning systems are the largest energy consumers in modern buildings, accounting for over 50% of a building's total energy use [1,2]. The central air-conditioning cooling source system includes chillers, chilled-water pumps, cooling-water pumps, and cooling towers, among others. Its energy consumption accounts for more than half of the total energy consumption of the central air-conditioning system. Therefore, optimizing the operation of the central air-conditioning cooling source system is the key to energy conservation in buildings.

Scholars both at home and abroad have conducted extensive research on energy-saving and efficiency improvement of central air-conditioning cooling source systems. Genetic algorithm and particle swarm optimization are the two most widely used intelligent algorithms. Chan et al. [3] used artificial neural networks to predict outdoor temperature, load, and chiller energy consumption, and to optimize the operating parameters of refrigeration plant equipment based on the particle swarm optimization algorithm. After optimization, the system performance coefficient increased by about 8.6%, and the total energy consumption decreased by about 7.9%. Wang et al. [4] proposed a hybrid optimization method combining PSO with association rules to distribute chiller loads, achieving a 12.5% reduction in total chiller energy consumption during the test day. Parouha & Verma [5] provide a systematic overview of recent advances in differential evolution and PSO, demonstrating that hybrid variants achieve faster convergence and better global search when optimizing HVAC systems. Similarly, Tian et al. [6] applied an improved DQN to central air-conditioning systems and achieved stable energy-saving control under large action-space conditions. The improved sparrow search algorithm was applied to the optimal chiller loading problem, and its advantages, such as greater energy savings, faster convergence, shorter running time, and improved robustness, were verified through experimental simulations [7]. The improved flower pollination algorithm is recommended for optimal chiller loading in multi-chiller air conditioning systems, particularly when higher robustness is required, even with a slightly reduced convergence rate [8].

In summary, optimizing the operating parameters of the central air-conditioning cooling source system has profound significance for energy conservation in buildings. Among emerging swarm intelligence algorithms, the whale optimization algorithm [9] has a simple structure, easy implementation, fewer parameters, and strong search ability. It is a powerful tool for solving complex engineering optimization problems and has been widely applied in fields such as low-carbon economic operation optimization of integrated energy systems [10], microgrid optimization [11], and bi-level planning of distributed power sources [12]. Therefore, this paper takes the central air-conditioning cooling source system of a super high-rise building in Guangzhou with an asymmetric configuration (1 small + 4 large chillers) as the object, constructs a mixed-integer optimization model considering load-equipment capacity matching, proposes an improved WOA strategy with collaborative encoding of discrete and continuous variables, verifies the energy-saving effect through three load scenarios of low, medium, and high, and further conducts parameter sensitivity analysis, aiming to distill implementable operation optimization methods and guidelines.

The remainder of this paper is organized as follows: Section 2 presents the models and problem formulation; Section 3 details the WOA with hybrid encoding; Section 4 offers the load analysis and case studies; Section 5 provides the algorithm comparisons and sensitivity analysis; and Section 6 concludes the work.

2. System Modeling and Problem Formulation

2.1. Project Overview

The research object is a 530 m tall super high-rise building in Guangzhou, with a total floor area of approximately 350,000 m^2. It functions as a large-scale international complex integrating high-end shopping malls, Grade A office spaces, serviced apartments, and luxury hotels.

As shown in Figure 1, this central air-conditioning cold-source system consists of 5 chillers, 5 chilled-water pumps, 5 cooling-water pumps, and 9 cooling towers. The main technical parameters for each piece of equipment (manufacturers: York for chillers, Wilo for pumps, EVAPCO for cooling towers) are summarized in Table 1.

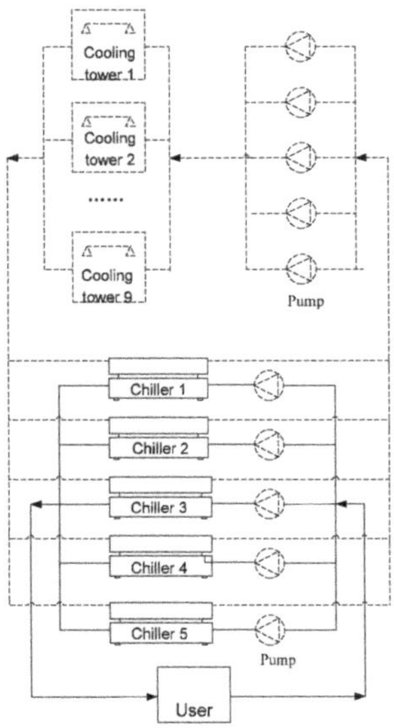

Figure 1. Schematic diagram of the central air-conditioning cold source system.

Table 1. Main technical parameters of each equipment in the cold source system.

Equipment Type	Key Technical Parameters	Data Sources
Chiller 1	Chilled water: 7 °C (inlet), 12 °C (outlet); Cooling water: 31 °C (inlet), 36 °C (outlet); Capacity: 4922 kW; Power: 859.4 kW; IPLV: 9.775; EER: 5.727; Chilled water flow: 844.9 m^3/h; Cooling water flow: 988.9 m^3/h; Chilled water pressure drop: 65.6 kPa; Cooling water pressure drop: 63.9 kPa	York Manufacturer's Technical Manual for Chiller Model YKQ5Q1K25DBH
Chillers 2–5	Chilled water: 7 °C (inlet), 12 °C (outlet); Cooling water: 31 °C (inlet), 36 °C (outlet); Capacity: 8439 kW; Power: 1377 kW; IPLV: 7.504; EER: 6.13; Chilled water flow: 1448.6 m^3/h; Cooling water flow: 1688.4 m^3/h; Chilled water pressure drop: 78.4 kPa; Cooling water pressure drop: 71.4 kPa	York Manufacturer's Technical Manual for Chiller Model YKV8W6K45DJH
Chilled Water Pump 1	Power: 110 kW; Head: 38 m; Flow rate: 847 m^3/h	WiLO Manufacturer's Technical Manual for Pump Model SCH250/360-110/4
Chilled Water Pumps 2–5	Power: 200 kW; Head: 38 m; Flow rate: 1452 m^3/h	WiLO Manufacturer's Technical Manual for Pump Model SCH300/430-200/4
Cooling Water Pump 1	Power: 110 kW; Head: 26 m; Flow rate: 1094 m^3/h	WiLO Manufacturer's Technical Manual for Pump Model SCH250/380-110/4
Cooling Water Pumps 2–5	Power: 200 kW; Head: 26 m; Flow rate: 1861.75 m^3/h	WiLO Manufacturer's Technical Manual for Pump Model SCH300/430-200/4
Cooling Towers 1–9	Power: 37 kW; Hot water temperature: 36 °C; Cold water temperature: 31 °C; Flow rate: 866 m^3/h; Wet-bulb temperature: 28 °C	EVAPCO Manufacturer's Technical Manual for Cooling Tower Model LCCM-N-800

2.2. Equipment Energy Consumption Models

2.2.1. Chiller Model

Empirical models for chiller efficiency include physics-based models such as the Gordon-NG model [13] and the Lee model [14], as well as statistically based models like the quadratic linear BQ model [15] and multivariate polynomial regression (MP) model [16].

To address the limited accuracy of traditional empirical models, this study constructs a multi-layer perceptron (MLP) regression model with three hidden layers, using over 20,000 sets of actual operating data provided by York Chiller Manufacturer. The data was collected from 1 May to 31 October 2024, with a sampling interval of 10 min per record, covering the chiller's operating states across different seasons and time periods. For outlier filtering, a combined method of "physical threshold screening + Isolation Forest algorithm" was adopted, and 15,892 valid data sets were finally retained for model training.

This MLP model adopts a hierarchically decreasing network structure of (64-32-16). The input layer includes five key operating parameters: chilled water flow rate, cooling water flow rate, chilled water supply temperature, load rate, and cooling water supply temperature. Each hidden layer uses the ReLU activation function, and the Adam optimization algorithm is employed for gradient descent training. The training results show that the neural network model achieves a coefficient of determination (R^2) of 0.9939, a root mean square error (RMSE) of 0.0822, a mean absolute error (MAE) of 0.0326, and a mean absolute percentage error (MAPE) of only 0.43% on the test set—significantly outperforming the traditional multivariate polynomial regression (MP) model. Furthermore, the difference in R^2 between the training and test sets is less than 0.003, indicating that the model has excellent generalization ability without overfitting, thus providing a high-fidelity energy consumption benchmark for subsequent optimization. The performance evaluation of the model can be referred to in Figure 2.

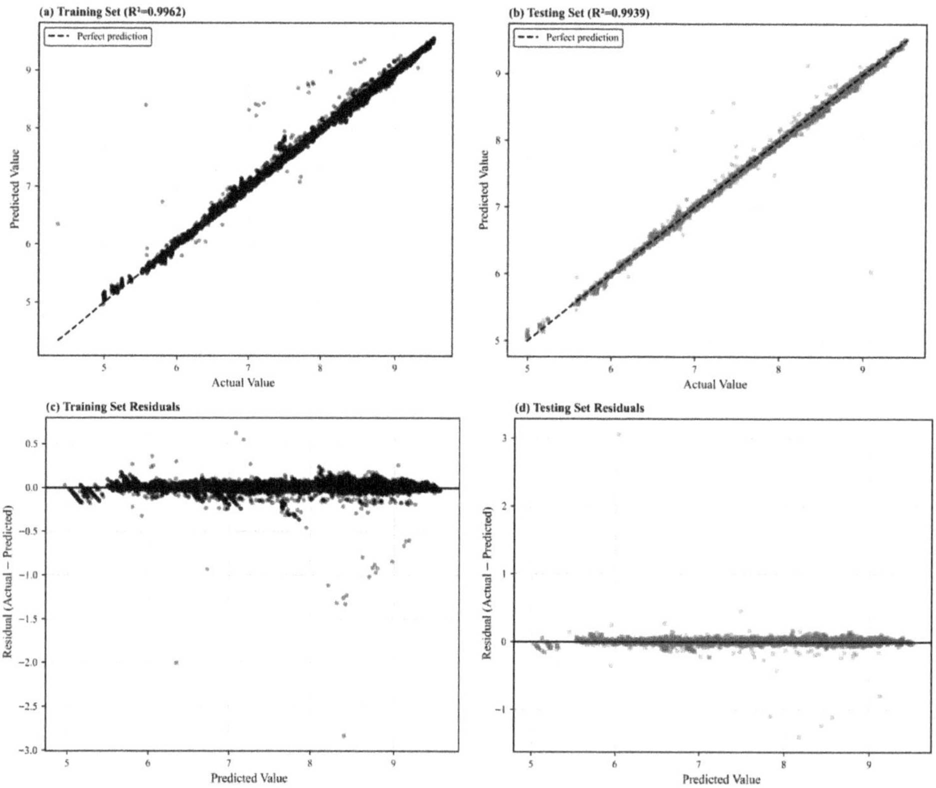

Figure 2. Performance evaluation of the MLP regression model. (**a**) Training set predictions vs. actual values (R^2 = 0.9962). (**b**) Testing set predictions vs. actual values (R^2 = 0.9939). (**c**) Training set residual distribution. (**d**) Testing set residual distribution. The results demonstrate excellent prediction accuracy and generalization capability with no systematic bias.

2.2.2. Pump Model

The quadratic pump-flow relationship is based on the experimental curve provided by Braun [17]. The flow-head characteristic curve of the pump is fitted using the least-squares method based on the manufacturer's operating data. The energy consumption models of the chilled water pump and the cooling water pump are shown in Equation (1).

$$W_p = b_0 + b_1 Q + b_2 Q^2 \quad (1)$$

where Q is the pump flow rate; b_0~b_2 are regression coefficients, and their specific values corresponding to different pump types are listed in Table 2.

Table 2. Polynomial coefficient of pump performance curves.

Pump Type	Pump ID	Rated Power (kW)	Rated Flow (m^3/h)	Rated Head (m)	b_0	b_1	b_2
Chilled Water Pump	CHWP-1	110	847	38	46.13	0.001580	−0.00001537
Chilled Water Pump	CHWP-2	200	1452	38	49.15	0.000220	−0.00000545
Cooling Water Pump	CWP-1	110	1094	26	36.00	0.002843	−0.00001103
Cooling Water Pump	CWP-2	200	1862	26	45.00	0.001200	−0.00000600

2.2.3. Cooling Tower Model

The energy consumption of cooling tower fans has traditionally relied on the Braun model [17], which is a semi-empirical model for cooling tower performance calculation. Although it can predict heat rejection and performance by combining thermodynamic equilibrium and empirical parameters, it fails to establish a coupling relationship between the cooling tower approach, chiller energy consumption, and overall system efficiency. Under the same cooling load, it is difficult to reflect the impact of approach differences on system optimization.

To address this, this paper uses the 21,369 sets of cooling tower flow conversion tables provided by the Cooling Technology Institute (CTI) for arbitrary and standard operating conditions, from which 20,695 valid data sets were selected to construct a Gradient Boosting Regressor model. The model is configured with 200 decision trees, a learning rate of 0.1, and a maximum depth of 4. The test results show that the coefficient of determination R^2 reaches 0.9977, the RMSE is 0.1034, the MAE is 0.0485, and the MAPE is only 4.65% (Figure 3), indicating extremely high prediction accuracy. Feature importance analysis further indicates that the temperature difference (RANGE) has the highest impact on the model's predicted values (59.25%), while the approach (APPROCH), as a key input parameter, can be precisely captured by the model for its coupling effect on system efficiency optimization, thus overcoming the limitations of the traditional Braun model.

The energy consumption calculation process for cooling towers is as follows:

First, a gradient boosting regression tree is used to establish the mapping relationship of the performance coefficient:

$$\eta = f(x) = \mathrm{GBR}\left(\Delta T_{range}, T_{approach}, T_{wb}\right) \quad (2)$$

The input feature vector is defined as follows:

$$x = \left[\Delta T_{range}, T_{approach}, T_{wb}\right]^T \quad (3)$$

where ΔT_{range} is the temperature difference between the inlet and outlet of the cooling water, $T_{approach}$ is the cooling tower approach, and T_{wb} is the ambient wet-bulb temperature.

The actual energy consumption of the cooling tower can then be calculated based on the performance coefficient correction method:

$$E_{tower} = C * P_{rated} * \frac{\eta}{\eta_0} * \frac{Q_{actual}}{Q_{rated}} \tag{4}$$

In the formula, η is the cooling tower performance coefficient under the current operating conditions, predicted by the gradient boosting regression model based on the cooling water temperature difference (ΔT_{range}), approach temperature ($T_{approach}$), and ambient wet-bulb temperature (T_{wb}); η_0 is the performance coefficient under the baseline operating conditions, i.e., the performance coefficient when ΔT_{range}, $T_{approach}$, and T_{wb} are all at the rated parameters of the equipment, calculated by the same model; the constant C = 0.871 is a unit conversion coefficient used to unify the dimensions of the performance coefficient and power parameters, and its value is derived from the unit conversion specifications under the CTI standard conditions. The physical meanings and calculation methods of the parameters in the equation are shown in Table 3:

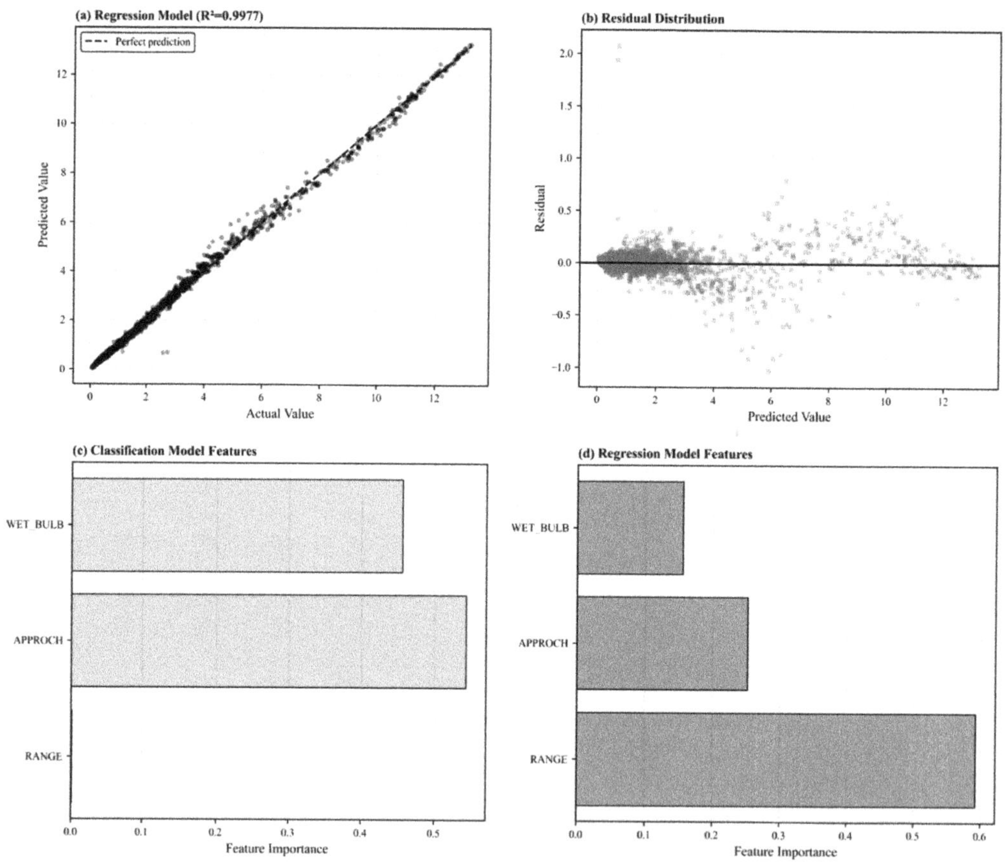

Figure 3. Evaluation of machine learning models for cooling tower performance prediction. (**a**) Regression model: predicted vs. actual values (R^2 = 0.9977, RMSE = 0.1034). (**b**) Residual distribution of regression predictions. (**c**) Feature importance ranking for classification model (missing value detection). (**d**) Feature importance ranking for regression model (performance coefficient prediction).

Table 3. Energy consumption parameters of cooling towers.

Parameter	Meaning	Calculation/Value Method
P_{rated}	Rated power of the cooling tower (kW)	Equipment nameplate parameter
Q_{rated}	Rated water flow of the cooling tower (m^3/h)	Equipment nameplate parameter
Q_{actual}	Actual water flow of the cooling tower (m^3/h)	Actual operating parameter
η	Performance coefficient under current operating conditions	$\eta = f(\left(\Delta T_{range}, T_{approach}, T_{wb}\right)$
η_0	Performance coefficient under base operating conditions	$\eta_0 = f(\left(\Delta T_{range,0}, T_{approach,0}, T_{wb,0}\right)$
C	Unit conversion coefficient	C = 0.871

2.3. Problem Model

2.3.1. Optimization Objective

For the central air-conditioning cold-source system, the optimization objective is to achieve the optimal energy efficiency of the cold-source system, i.e., to maximize the EER of the cold-source system. The EER is defined as the ratio of the total cooling capacity of the system to the total energy consumption of the system, as shown in Equation (5):

$$EER = \frac{Q_{c,total}}{W_{ch} + W_p + P_{fan}} \tag{5}$$

where Q_{total} is the total cooling capacity of the system; W_{ch}, W_p, and P_{fan} are the power consumptions of the chillers, pumps, and cooling towers, respectively. Therefore, the optimization objective can be described as follows:

$$\max EER = \max\left(\frac{Q_{c,total}}{W_{ch} + W_p + P_{fan}}\right) \tag{6}$$

2.3.2. Optimization Variables and Constraints

The optimization variables include the status and operating parameters of the following equipment, which are adjusted to enhance the overall system efficiency. The optimization process must adhere to multiple physical and operational constraints. The flow rates of the chillers, chilled water pumps, cooling water pumps, and cooling towers must operate within their minimum and maximum design flow ranges and cannot exceed their physical limitations. Additionally, the approach of the cooling tower, limited by its own heat dissipation capacity, must be maintained within a reasonable range. It cannot be too small, exceeding the cooling tower's capability, nor too large, as this would reduce the overall cooling effect of the system. Moreover, the load on the chillers and pumps must not fall below their minimum load requirements to ensure they operate efficiently. Overall, the operating status of all equipment, flow distribution, and the approach of the cooling tower must comply with the system's design capabilities and actual operational requirements. In summary, the constraints are specifically described as follows:

$$\begin{cases} T_{approach,\min} < \Delta T_{approach} \\ f_{ct,\min} < f_{ct} < f_{ct,\max} \\ f_{chiller,\min} < f_{chiller} < f_{chiller,\max} \\ f_{chw\_pump,\min} < f_{chw\_pump} < f_{chw\_pump,\max} \\ f_{cw\_pump,\min} < f_{cw\_pump} < f_{cw\_pump,\max} \\ 0 \le N_{ct} \le N_{ct,\max},\ N_{ct} \in Z \\ 0 \le N_{chiller} \le N_{chiller,\max},\ N_{chiller} \in Z \\ 0 \le N_{chw\_pump} \le N_{chw\_pump,\max},\ N_{chw\_pump} \in Z \\ 0 \le N_{cw\_pump} \le N_{cw\_pump,\max},\ N_{cw\_pump} \in Z \end{cases} \tag{7}$$

where: Z is the set of integers.

The optimization variables and specific constraints are detailed in Table 4.

Table 4. Optimization variables and constraints.

Equipment Category	Variable Description	Variable Symbol	Unit	Constraint	Remarks
Cooling Tower	Cooling Tower Approach temperature	$\Delta T_{approach}$	°C	2.0~8.0	Difference between the cooling water outlet temperature and the wet-bulb temperature
	Number of Operating Cooling Towers	N_{ct}	Units	1~9	Must match the number of chillers and cooling water pumps in operation
	Flow Distribution of Cooling Towers	f_{ct}	-	0.3~1 (Proportion of single-unit flow to rated flow)	Single-unit flow must not be lower than 30% of the rated flow (to avoid uneven heat dissipation due to low flow)
Chiller	Number of Operating Chillers	$N_{chiller}$	Units	1~5	Must meet the total cooling load demand of the system
	Load Distribution of Operating Chillers	$f_{chiller}$	-	0.15~1 (Proportion of single-unit load to rated load)	Single-unit load must not be lower than 15% of the rated capacity (to prevent frequent start-stop of equipment)
Chilled Water Pump	Number of Operating Chilled Water Pumps	N_{chw_pump}	Units	1~5	Must be adapted to the number of chillers in operation
	Operating Frequency of Chilled Water Pumps	f_{chw_pump}	Hz	30.0~50.0	Set according to the chilled water flow demand, the lower limit is the minimum safe operating frequency of the pump (to avoid cavitation), upper limit is the rated frequency
Cooling Water Pump	Number of Operating Cooling Water Pumps	N_{cw_pump}	Units	1~5	Must be adapted to the number of chillers in operation
	Operating Frequency of Cooling Water Pumps	f_{cw_pump}	Hz	30.0~50.0	Set according to the cooling water flow demand, the lower limit is the minimum safe operating frequency of the pump (to avoid cavitation), upper limit is the rated frequency

Note: Among the 29 optimization variables, 9 are discrete on/off decisions and 20 are continuous operating parameters.

3. Algorithm Design

3.1. Application of the Whale Optimization Algorithm

The Whale Optimization Algorithm (WOA) is a new swarm intelligence optimization algorithm that simulates the hunting behavior of humpback whales. In the WOA, each whale's position represents a feasible solution. The hunting behavior of whales is divided into three stages: encircling prey, bubble-net attacking, and searching for prey. The optimization process is shown in Figure 4. In the optimization of central air-conditioning cooling source systems, the WOA can handle both continuous and discrete variables simultaneously. Continuous variables include chiller load, pump frequency, cooling tower flow, and approach temperature, while discrete variables include the equipment's on/off states. Compared with GA and PSO algorithms, WOA does not require manual adjustment of inertia weights. It relies on the convergence factor A to adaptively balance global and local search, reducing the parameter-tuning burden [9]. Its "encircling prey" and "bubble-net attacking" mechanisms effectively suppress premature convergence. Moreover, WOA uses a "threshold truncation" approach for discrete variables, eliminating the need for binary encoding required by GA, making it more concise and efficient and improving computational efficiency by about 20%.

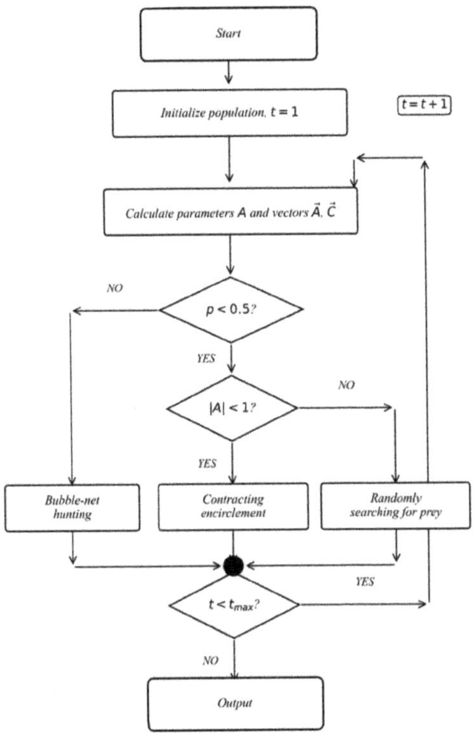

Figure 4. Flowchart of the whale optimization algorithm.

3.2. Initialization and Solution Update

The initialization process is to generate the initial population, which is carried out according to the following steps:

For the load distribution of chillers, random numbers are generated based on the minimum load limit and the maximum load for each chiller. If the total load during initialization is not zero (i.e., at least one chiller is allocated a load), the load is adjusted to the total cooling load through proportional scaling. For the cooling tower approach temperature, initial values are randomly generated within a reasonable range (between the minimum approach temperature and 8 °C).

In this optimization, the update of solutions follows the mechanism of WOA. For each whale, the update equation is as follows:

$$X(t+1) = X_{rand} - A \times D \tag{8}$$

When $p < 0.5$ and $|A| \geq 1$, the update equation is as follows:

$$X(t+1) = X^*(t) - A \times D \tag{9}$$

When $p \geq 0.5$, perform a search using the spiral search equation as follows:

$$X(t+1) = D^* \times e^{bl} \times cos(2\pi l) + X^*(t) \tag{10}$$

where C, b, and l are random coefficients.

3.3. Constraint Control and Handling of Discrete Variables

During the optimization process, WOA needs to handle a variety of constraints to ensure the physical feasibility of the solution and the safe operation of the equipment. The constraint handling strategies adopted in this study are as follows:

3.3.1. Chiller Load Constraint Adjustment

A combination of boundary truncation and threshold strategies is used. For the i-th chiller, the feasible range of load Q_i (subscript i denotes the i-th chiller) is $[Q_{i,min}, Q_{i,max}]$. Here $Q_{i,min} = \alpha \cdot Q_{i,rated}$ (where α is the minimum load rate of the chiller, set to 0.15 based on the manufacturer's technical parameters), and $Q_{i,max} = \alpha \cdot Q_{i,rated}$ (rated cooling capacity of the i-th chiller). During the solution update process:

- If $Q_i < Q_{i,min}$, set $Q_i = 0$ (shut down the i-th chiller);
- If $Q_i > Q_{i,max}$, (limit to the rated capacity of the i-th chiller).

3.3.2. Cooling Tower Approach Constraint Adjustment

The feasible range of the approach is determined by two conditions:

1. The physical lower bound Amin based on the cooling tower performance model (predicted by the gradient boosting regression model using the current wet-bulb temperature and flow ratio);
2. The safety constraint that the cooling water supply temperature must not be lower than 18 °C, i.e., $T_{wb} + A \geq 18$;
3. The upper limit constraint of the cooling water supply temperature of 32 °C, i.e., $T_{wb} + A \leq 32$ (an upper limit given from the perspective of system safe operation). Combining the above conditions, the approach boundary is $\in$[max(Amin, 18 $-T_{wb}$), 32 $-$ T_{wb}]. If the updated approach exceeds this range, it is truncated to the feasible domain through a calculation function.

3.3.3. Total Cooling Load Balance Constraint

A two-stage adjustment mechanism is used to ensure that the total load precisely matches the target value Q_{target}.

Stage 1: Scale the load of all operating chillers proportionally. The scaling factor is defined as follows:

$$\lambda = \frac{Q_{target}}{\sum_{i=1}^{n} Q_i} \tag{11}$$

where Q_{target} is the total cooling load demand of the system; n is the number of operating chillers; and Q_i is the initial load of the i-th operating chiller.

Stage 2: If the total load after scaling deviates from Q_{target}, eliminate the deviation by fine-tuning each operating chiller's load one by one. Prioritize chillers with larger load margins (i.e., the difference between $Q_{i,max}$ and current Q_i, or between current Q_i and $Q_{i,min}$) to ensure the final total load satisfies:

Specific adjustment strategies:

If the total load exceeds the target, reduce the load of each unit one by one, with each unit being reduced to its minimum load Q_i, min at most;

If the total load is below the target, increase the load of each unit one by one, with each unit being increased to its rated capacity Q_i, max at most.

3.3.4. Discrete Variable (On/Off State) Processing

A "threshold truncation + continuous relaxation" strategy is adopted. During optimization, discrete variables are relaxed into continuous variables (load $Q_i \in [0, Q_{i,max}]$), and the discrete characteristics are restored through threshold judgment:

When $Q_i > 0$, the on/off state of the unit $s_i = 1$ (on);

When $Q = 0$, $s_i = 0$ (off).

This method avoids the combinatorial explosion problem of discrete optimization while ensuring the physical feasibility of the solution. In actual implementation, the chiller is automatically shut down when its load is below the minimum load threshold, reflecting the principle of economical operation.

3.3.5. Convergence Control

An early stopping mechanism is introduced to improve algorithm efficiency and avoid ineffective iterations. The patience parameter is set to patience = 10, and the tolerance threshold is tol = 10^{-6}. If the best improvement in the objective function value (i.e., the difference between the current optimal value and the optimal value from 10 generations ago) is less than tol for 10 consecutive iterations, the search is deemed stagnant, and the algorithm is terminated early.

To ensure reproducibility and consistency, the population size used in this study was set to 50, and the maximum number of iterations was set to 100. These parameters were kept constant across all optimization runs to ensure the stability and robustness of the WOA algorithm.

4. Optimization Analysis

4.1. Building Cooling Load Characteristics

The cooling load characteristics of the office building on the user side are calculated using the DeST software. The simulation is conducted using DeST v3.0 (Designer Simulation Toolkit), developed by Tsinghua University, Beijing, China. The simulation followed the procedure described by Yan et al. [18]. The settings of parameters such as the thermophysical properties of the building envelope, indoor heat gain, and indoor air temperature and humidity comply with the provisions of Article 3.3.1 "Thermal Performance Limits of Building Envelopes in Hot-Summer and Warm-Winter Areas" and Chapter 4 "General Provisions for Heating, Ventilation, and Air Conditioning Design" in the *Design Standard for Energy Efficiency of Public Buildings (GB 50189-2015)* [19]. The indoor air temperature setpoint is 25 °C during operating hours, and the cooling load is weighted averaged for each day of every month to derive the monthly weighted average cooling load characteristic distribution of the building, as shown in Figure 5.

As depicted in Figure 5, the cooling load characteristic in June is similar to that of July and August; May and September are similar; April and October are similar; and there is little difference between November, December, and January.

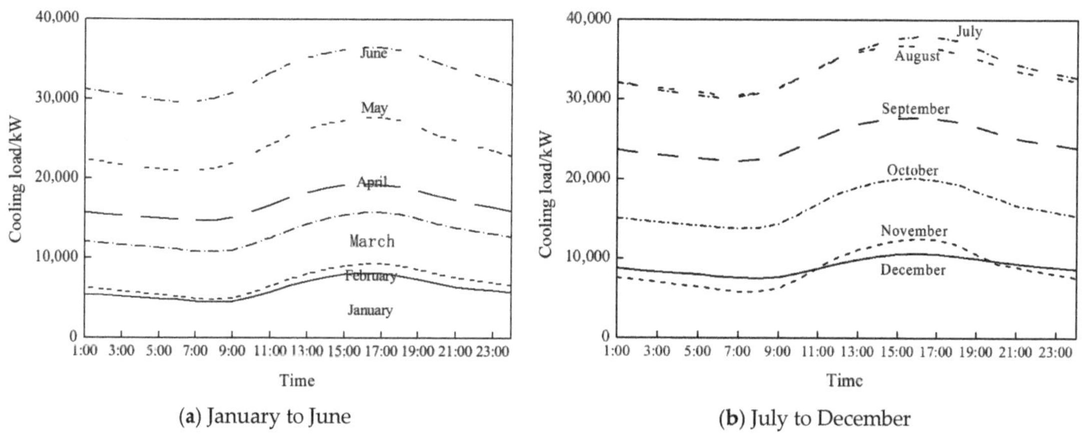

Figure 5. Annual cooling load characteristics of the building.

4.2. Comparison Before and After Optimization

Based on the system model, problem formulation, and algorithm design, the optimization problem for the central air-conditioning cooling source system is implemented using Python programming. This optimization system is realized using Python 3.12, with major dependencies including NumPy 2.3.3, pandas 2.3.3, and SciPy 1.16.2. The source code and related data can be obtained from the corresponding author. Three representative operating conditions (as shown in Table 5) are selected for detailed analysis to evaluate the effectiveness of WOA in improving system efficiency.

Table 5. Summary comparison of three operating conditions.

Item	Operating Condition 1	Operating Condition 2	Operating Condition 3
Load (kW)	11,592	22,702	34,059
Load Percentage (%)	30.9	60.6	90.9
Dry-Bulb Temperature (°C)	16.3	26.4	30.1
Wet-Bulb Temperature (°C)	14.8	24.0	26.4

4.2.1. Analysis of Operating Condition 1 (Low Load Scenario)

Under Operating Condition 1, the system consumed 12.91% less energy after optimization. It dropped from 2094.5 kWh to 1824.0 kWh. The Energy Efficiency Ratio (EER) increased from 5.53 to 6.36, a 15.01% improvement. The chiller used the most energy. The chilled water pump ranked second. The cooling water pump came next. The cooling tower had the smallest share. After optimization, the energy consumption of the chiller, chilled water pump, and cooling water pump was reduced by 7.23%, 45.23%, and 34.84%, respectively, while the cooling tower's energy consumption increased by 0.19%. This shows there were trade-offs in the optimization. Under this low-load operating condition, the optimization algorithm effectively adjusted the operating parameters of the equipment, achieving significant energy savings while ensuring the required cooling capacity. The detailed energy-consumption breakdown is illustrated in Figure 6, and the corresponding operating parameters before and after optimization are listed in Table 6.

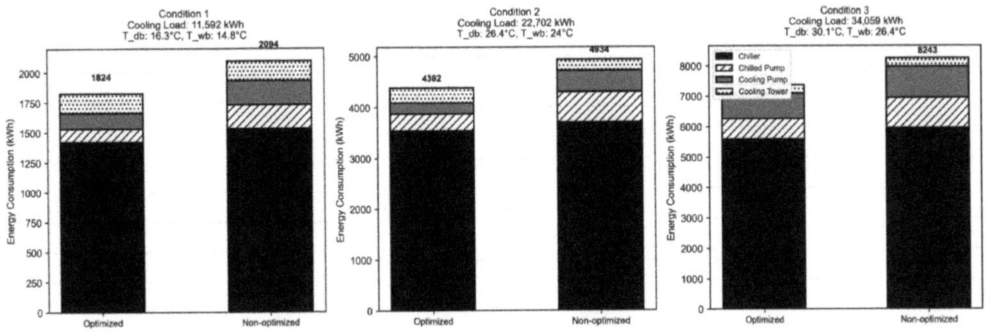

Figure 6. Energy consumption comparison before and after optimization.

Table 6. Operating parameters before and after optimization.

Condition	Parameter	Before Optimization	After Optimization
Case 1	Chiller Load Distribution (kW)	[4270.0, 7321.0, 0.0, 0.0, 0.0]	[3172.2, 0.0, 0.0, 0.0, 8419.8]
Case 2	Chiller Load Distribution (kW)	[0.0, 7567.0, 7567.0, 7567.0, 0.0]	[2593.4, 0.0, 6798.2, 6757.3, 6553.1]
Case 3	Chiller Load Distribution (kW)	[4334.0, 7431.0, 7431.0, 7431.0, 7431.0]	[2971.3, 7775.1, 7771.3, 7776.5, 7764.7]
Case 1	Chilled Water Pump Frequency (Hz)	[48.2, 47.4, 0.0, 0.0, 0.0]	[30.4, 30.4, 0, 0, 0]
Case 2	Chilled Water Pump Frequency (Hz)	[0.0, 48.0, 48.0, 48.0, 0.0]	[0, 36.4, 36.4, 36.4, 0]
Case 3	Chilled Water Pump Frequency (Hz)	[50.0, 50.0, 50.0, 50.0, 50.0]	[42.9, 42.9, 42.9, 42.9, 42.0]
Case 1	Cooling Water Pump Frequency (Hz)	[40.0, 40.0, 0.0, 0.0, 0.0]	[35.3, 35.3, 0, 0, 0]
Case 2	Cooling Water Pump Frequency (Hz)	[0.0, 45.0, 45.0, 45.0, 0.0]	[41.1, 41.1, 41.1, 0, 0]
Case 3	Cooling Water Pump Frequency (Hz)	[50.0, 50.0, 50.0, 50.0, 50.0]	[0, 43.5, 43.5, 43.5, 43.5]
Case 1	Cooling Tower Flow Distribution (m^3/h)	[452.0, 452.0, 452.0, 452.0, 452.0, 0.0, 0.0, 0.0, 0.0]	[453.0, 453.0, 453.0, 453.0, 453.0, 0, 0, 0, 0]
Case 2	Cooling Tower Flow Distribution (m^3/h)	[495.0, 495.0, 495.0, 495.0, 495.0, 495.0, 495.0, 495.0, 495.0]	[494.0, 494.0, 494.0, 494.0, 494.0, 494.0, 494.0, 494.0, 494.0]
Case 3	Cooling Tower Flow Distribution (m^3/h)	[742.0, 742.0, 742.0, 742.0, 742.0, 742.0, 742.0, 742.0, 742.0]	[742.0, 742.0, 742.0, 742.0, 742.0, 742.0, 742.0, 742.0, 742.0]
Case 1	Approach Temperature (°C)	4	3.5
Case 2	Approach Temperature (°C)	4	2.5
Case 3	Approach Temperature (°C)	4	3.54

Note: [x_1, x_2, x_3, ..., x_i] respectively represent the capacity of equipment 1 to i, and 0 indicates "off".

4.2.2. Analysis of Operating Condition 2 (Medium Load Scenario)

Under Operating Condition 2, the system also demonstrated strong optimization potential. The total energy consumption was reduced by 11.18%, from 4933.8 kWh to 4382.0 kWh. The EER increased from 4.60 to 5.18, a 12.61% improvement. In terms of individual equipment energy consumption, the chiller, chilled water pump, and cooling water pump achieved energy savings of 4.35%, 44.98%, and 47.87%, respectively. During the optimization of Operating Condition 2 under medium load, the cooling tower's energy consumption increased significantly, mainly to balance the efficient operation of the chiller and pumps (such as matching equipment load distribution and ensuring heat dissipation requirements). By appropriately increasing the operating intensity of the cooling tower to support overall system efficiency improvement, this was a reasonable trade-off in the optimization process.

4.2.3. Analysis of Operating Condition 3 (High Load Scenario)

Under Operating Condition 3, which is a high-load scenario, the energy consumption was reduced by 10.58%, from 8242.7 kWh to 7371.0 kWh. The EER increased from 4.13 to 4.62, a 11.86% improvement. In terms of individual equipment, the chiller, chilled water pump, and cooling water pump achieved energy savings of 5.92%, 33.57%, and 18.07%, respectively. The cooling tower's energy consumption remained essentially unchanged. Under high-load operating conditions, although the optimization space was relatively limited due to the need for most equipment to operate at higher capacities, the Whale Optimization Algorithm still achieved significant energy savings by finely adjusting operating parameters and load distribution among chillers.

4.3. Comparative Analysis of the Three Operating Conditions

A comprehensive comparison of the three operating conditions reveals that the achieved energy savings rates are 12.91%, 11.18%, and 10.58%, respectively. This is illustrated in Figure 7 (energy consumption and saving rate) and Figure 8 (EER comparison). This indicates that the optimization algorithm is effective across different load levels, with the most significant energy-saving effects observed under low to medium load conditions.

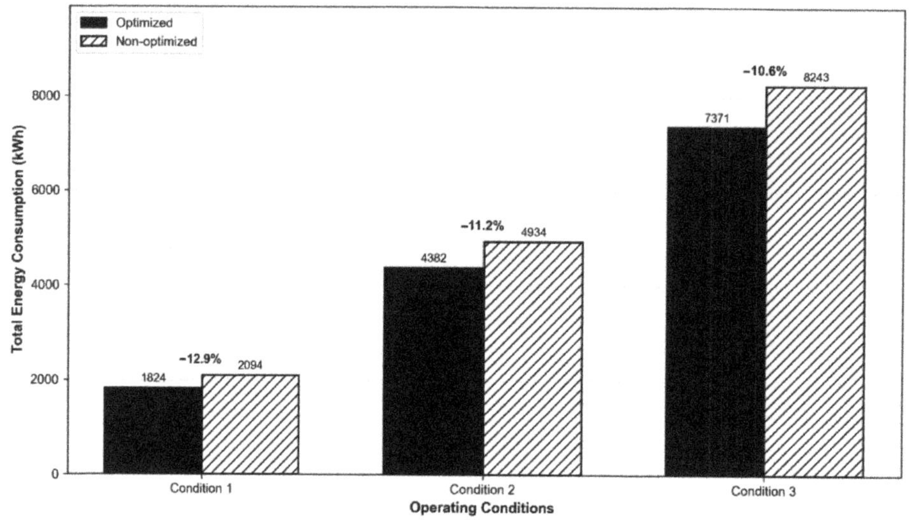

Figure 7. Total Energy consumption and saving rate.

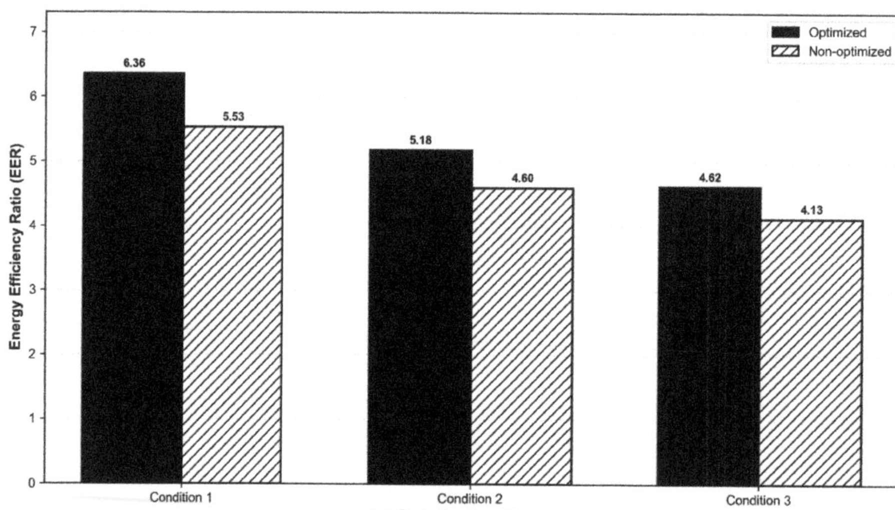

Figure 8. EER comparison before and after optimization.

The EER improvement percentages for the three operating conditions are 15.01%, 12.61%, and 11.86%, respectively. The EER decreases with higher outdoor temperatures, which aligns with thermodynamic principles. However, the relative improvement remains significant under all conditions.

The chilled water pumps and cooling water pumps demonstrate significant optimization potential, especially under medium and low load conditions, with energy savings reaching 44.98% and 47.87%, respectively. This highlights the importance of variable frequency drive control in achieving overall system optimization.

The research results indicate that under extremely low or high load conditions, the number of operating units before and after optimization tends to be the same, with the optimization space mainly limited to capacity adjustment. In contrast, under medium load conditions, there is optimization potential in both the number of units and their capacities, resulting in greater energy savings and EER improvements.

4.4. Annual Energy Consumption Comparison

Based on the monthly optimization results, the WOA algorithm shows good adaptability under different seasonal and load conditions. The average annual energy savings rate is 13.42%, with the EER increasing from 5.07 to 5.88, a 16.10% improvement. Specifically, the optimization effect is most significant under low-load conditions in winter, with an energy savings rate of 18.92% and a 23.3% EER improvement in December. Although the energy savings rate is relatively lower under high-load conditions in summer (12.01% in July), the absolute amount of energy saved is still considerable due to the high base energy consumption. The energy savings rate for each month remains between 8.59% and 18.92%, indicating that the algorithm can adaptively adjust equipment configuration strategies according to different operating conditions, achieving stable and reliable energy savings across the entire operating range. This demonstrates its good engineering application value, as summarized in Figure 9.

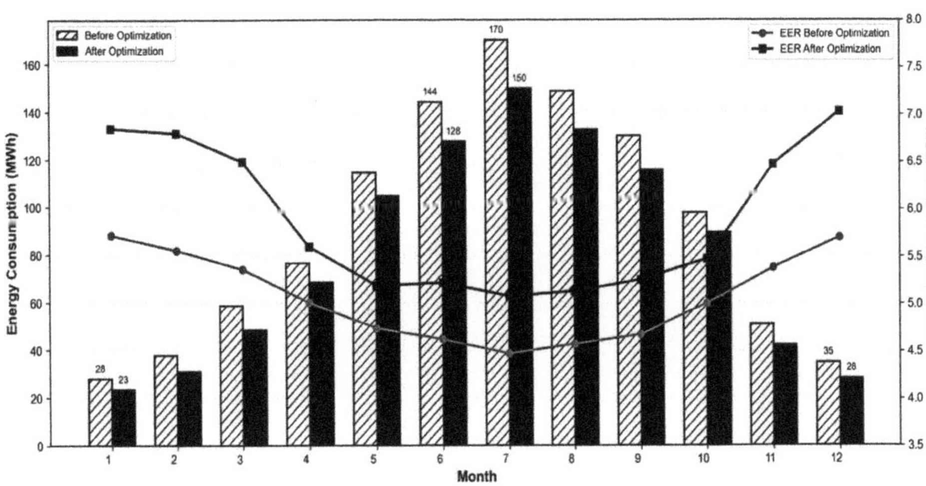

Figure 9. Comparison of monthly average weighted energy consumption and energy efficiency ratio before and after optimization.

5. Algorithm Comparison and Sensitivity Analysis

5.1. Optimization Algorithm Performance Comparison

To verify the effectiveness of WOA in the optimization of cooling source systems, WOA is compared with two classic algorithms, GA and PSO. All three algorithms use the same optimization objective (minimizing the total system energy consumption), constraints, and initial parameter settings to ensure a fair comparison.

The algorithm parameter settings are as follows: The population size is 50 for all algorithms, with a maximum of 100 iterations, convergence tolerance tol = 10^{-6}, and patience parameter patience = 10. For GA, tournament selection, single-point crossover (crossover probability 0.8), and Gaussian mutation (mutation probability 0.2) are used. For PSO, the inertia weight linearly decreases from 0.9 to 0.4, with learning factors c1 = c2 = 2.0. For WOA, the parameter linearly decreases from 2 to 0. To ensure statistical significance of the results, each algorithm is run 10 times under three typical operating conditions (low load, medium load, and high load), with the average value taken as the final result.

Table 7 shows the optimization performance comparison of the three algorithms under the three typical operating conditions. In terms of optimization accuracy, all three algorithms converge to essentially the same optimal value under each condition, indicating that WOA has global search capabilities comparable to GA and PSO. From the perspective of computational efficiency, WOA achieves the shortest average computation time of 225.20 s, which is 50.6% faster than GA and 11.0% faster than PSO, demonstrating a higher convergence speed. In terms of system EER, the optimized system EER values for the three algorithms under different conditions are 6.355, 5.183, and 4.621, respectively, representing improvements of 8.2%, 12.5%, and 15.3% over the pre-optimization values, thus validating the effectiveness of intelligent optimization methods in enhancing system efficiency.

Table 7. Performance comparison of three optimization algorithms.

Case	Algorithm	Load (kW)	Wet-Bulb Temperature (°C)	Total Energy Consumption (kW)	EER	Computation Time (s)
Case 1 (Low Load)	GA	11,592	14.8	1824	6.355	396.52
	PSO	11,592	14.8	1824	6.355	170.61
	WOA	11,592	14.8	1823	6.356	120.41
Case 2 (Medium Load)	GA	22,702	24.0	4380	5.183	458.10
	PSO	22,702	24.0	4384	5.178	150.92
	WOA	22,702	24.0	4380	5.183	146.95
Case 3 (High Load)	GA	34,059	26.4	7371	4.621	513.52
	PSO	34,059	26.4	7371	4.621	437.16
	WOA	34,059	26.4	7371	4.621	408.23

Figure 10 shows the convergence curve comparison of the three algorithms across the three operating conditions. As can be seen from the figure, although all algorithms can find the optimal solution, WOA performs best in terms of convergence speed, with an average of only 49.7 iterations required. This is 35% faster than PSO and 50% faster than GA. This makes WOA particularly suitable for real-time optimization applications in building energy management systems.

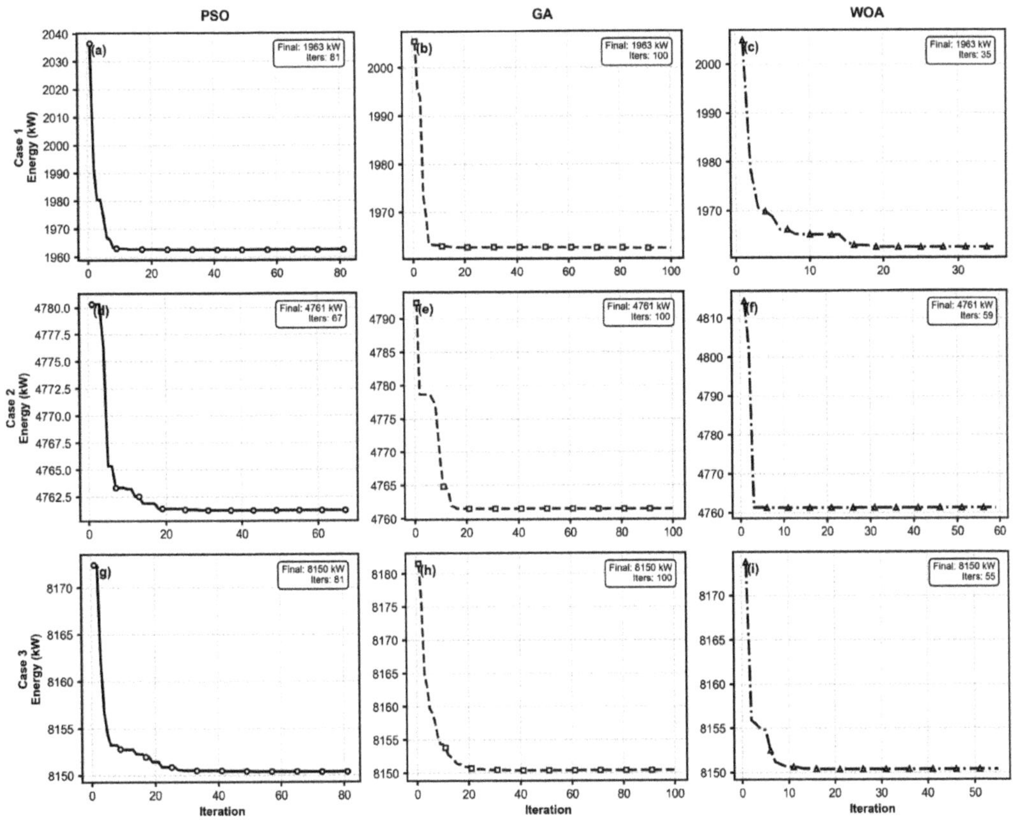

Figure 10. Convergence characteristics of PSO, GA, and WOA algorithms for three operating cases: (**a–c**) Case 1 with 11,592 kW cooling load at 14.8 °C wet-bulb temperature; (**d–f**) Case 2 with 22,702 kW cooling load at 24.0 °C wet-bulb temperature; (**g–i**) Case 3 with 34,059 kW cooling load at 26.4 °C wet-bulb temperature. WOA consistently demonstrated the fastest convergence speed across all operating conditions.

5.2. Stability Analysis of the Algorithms

Under operating condition 2, a stability assessment was conducted for WOA, GA, and PSO over 20 independent iterations. These 20 independent runs were initialized using different random seeds to ensure that each run started from a unique population. This approach allows for a comprehensive evaluation of the algorithms' stability under varying initial conditions. The results show that all three algorithms achieved a 100% success rate, with extremely small stability differences in terms of energy consumption: the average energy consumption was approximately 4761–4763 kW, and the standard deviation and coefficient of variation (CV) were extremely low (e.g., WOA CV $\approx$ 0.046%, PSO $\approx$ 0.014%, GA $\approx$ 0.000%). These differences fall within the acceptable range of engineering measurement errors and normal operational fluctuations and have a limited impact on overall decision-making. In contrast, the time dimension is more decisive in engineering applications: WOA has the lowest average time ($\approx$248.92 s) and the smallest time standard deviation ($\approx$78.46 s), demonstrating higher execution predictability and scheduling reliability; PSO is in the middle; GA has a slightly lower time CV, but its average time is much longer, and it exhibits larger fluctuations. These factors reduce its timeliness and stability. In summary, given that differences in energy consumption stability are negligible in engineering terms, WOA has significant advantages in key time stability indicators (mean and standard deviation), making it more suitable for scenarios requiring real-time and high-frequency operation. The detailed statistics are listed in Table 8.

Table 8. Comparative analysis of algorithm stability.

Algorithm	Success Rate (%)	Mean Energy (kW)	Std Energy (kW)	Min Energy (kW)	Max Energy (kW)	CV Energy (%)	Mean Time (s)	Std Time (s)	CV Time (%)
WOA	100	4762.7	2.17	4761	4770	0.046	248.92	78.46	31.52
GA	100	4761	0	4761	4761	0	386.71	112.31	29.04
PSO	100	4761.15	0.65	4761	4764	0.014	262.68	92.76	35.31

5.3. Sensitivity Analysis of Key Parameters

To evaluate the impact of system configuration and operating parameters on the optimization results, this study conducts a sensitivity analysis on five key parameters using the experimental scheme shown in Table 9: number of chillers, number of cooling towers, number of pumps, lower limit of pump frequency, and cooling tower approach temperature. All analyses are based on single-factor perturbation experiments under Operating Condition 2.

Table 9. Experimental scheme for sensitivity analysis.

Index	Parameter Type	Baseline Value	Range of Change	Step Size	Number of Experiment Groups	Control Variables
1	Number of Chillers	4 units	4 to 5 units	1 unit	2	Number of cooling towers, number of pumps, pump frequency, and approach temperature
2	Number of Cooling Towers	5 units	5 to 9 units	2 units	3	Number of chillers, number of pumps, pump frequency, approach temperature
3	Number of Pumps	4 units	4 to 5 units	1 unit	2	Number of chillers, number of cooling towers, pump frequency, and approach temperature
4	Lower Limit of Pump Frequency	70%	70% to 100%	5%	7	Number of chillers, number of cooling towers, number of pumps, approach temperature
5	Cooling Tower Approach Temperature	4 °C	2 to 6 °C	1 °C	5	Number of chillers, number of cooling towers, number of pumps, pump frequency

Table 10 shows the sensitivity coefficients for the five types of parameters. The results indicate that the number of pumps has the most significant impact on system energy consumption, with a sensitivity coefficient of 1.1125, indicating that pump configuration is the most critical factor affecting system efficiency. Increasing the number of pumps from 4 to 5 reduces total system energy consumption from 4876 kW to 4823 kW, a decrease of 1.1%. Adding more pumps lowers the load on each one. This allows them to run more efficiently. The efficiency gain offsets the extra energy used by the additional pump.

Table 10. Summary and ranking of sensitivity coefficients.

Rank	Parameter Type	Parameter Range	Change in Total Energy Consumption (kW)	Change in EER	Sensitivity Coefficient	Impact Level
1	Number of Pumps	4 → 5 units	−53.0	+0.04	1.1125	High
2	Lower Limit of Pump Frequency	70% → 100%	+340.0	−0.31	0.1957	Medium
3	Cooling Tower Approach Temperature	2 → 6 °C	+208.5	−0.20	0.0431	Medium
4	Number of Chillers	4 → 5 units	−35.0	+0.04	0.0325	Low
5	Number of Cooling Towers	5 → 9 units	+6.2	0.00	0.0023	Low

The sensitivity coefficient of the lower limit of pump frequency is 0.1957, ranking second. Raising the minimum frequency from 70% to 100% increases energy use. Consumption rises from 4823 kW to 5163 kW. That is a 7.0% increase. The EER also drops from 4.71 to 4.40. This verifies the importance of variable frequency control for energy savings: within the frequency range of 70–80%, system energy consumption remains relatively stable; however, when the frequency exceeds 85%, energy consumption rises rapidly because the pump power is proportional to the cube of the rotational speed, and high-frequency operation leads to a sharp increase in pump energy consumption.

The sensitivity coefficient of the cooling tower approach temperature is 0.0431. When the approach temperature increases from 2 °C to 6 °C, the total system energy consumption rises from 4747 kW to 4955 kW, an increase of 4.4%, and the EER decreases from 4.78 to 4.58. An increase in approach temperature results in a higher cooling-water supply temperature, which leads to a decrease in chiller COP. Unlike the passive influence of wet-bulb temperature, the approach temperature is an optimizable control variable, and the optimization algorithm tends to select a smaller approach temperature (within the equipment performance limits) to improve system efficiency.

The sensitivity coefficients for the number of chillers and the number of cooling towers are 0.0325 and 0.0023, respectively, indicating relatively minor impacts. This is because, within the load range of this study, the existing configuration already meets the operational requirements. Increasing the number of units further results in diminishing marginal benefits and may even lead to efficiency degradation due to low part-load ratios.

6. Conclusions

This study focuses on the "1 small + 4 large" asymmetric central air-conditioning cooling source system of a 530 m super high-rise building in Guangzhou, addressing the 30 min rolling scheduling demand of the BA system. It breaks through the optimization bottleneck of "mixed integers, asymmetry, and strong constraints," forming a complete technical chain of "modeling–algorithm–patterns–verification." The main conclusions and innovations are as follows:

For the first time, a three-hidden-layer MLP chiller model trained on 16,276 sets of actual measurements and a gradient-boosting integrated cooling tower model driven by 21,369 operating conditions are constructed, supplemented by a secondary polynomial model for variable frequency pumps based on fluid mechanics similarity laws. This achieves high-precision modeling of energy consumption across all equipment on the cooling-source side, providing a reliable benchmark for asymmetric configuration optimization.

A hybrid encoding of "threshold truncation + continuous relaxation" is proposed, embedding 29-dimensional discrete on/off and continuous parameters into WOA. A three-layer constraint repair mechanism of "boundary–load–approach temperature" is designed, with a single optimization round taking ≤225 s. Compared with GA and PSO, the speed is increased by 41.3% and 25.5%, respectively, and the energy consumption coefficient of variation is only 0.08%, meeting the minute-level real-time scheduling requirements of the BA system.

The optimal load-capacity matching pattern for asymmetric cooling-source clusters is revealed: under low-load conditions, the small unit is mainly operated. When the load rate is around 60%, the configuration is switched to "1 small + 2 large" units, and when the load rate reaches 90%, it is further expanded to "1 small + 4 large" for full-unit operation. In terms of sensitivity analysis, the number of chilled water pumps is the most sensitive parameter, with a 1.1% energy saving by adding one pump. The lower limit of pump frequency and the cooling tower approach temperature are the next most sensitive factors, while the impact of the number of chillers and cooling towers is <0.05. This provides a quantifiable decision-making sequence table for operation and maintenance.

After a full-year rolling optimization, the average EER of the system increases from 5.07 to 5.88 (+15.98%), with an annual energy savings rate of 8.59–18.92%. peak energy savings occur in winter under low-load conditions (18.92% in December), while the absolute value of energy savings is highest in summer under high-load conditions. The typical operating condition energy savings range from 10.58% to 12.91%.

In summary, the WOA optimization method proposed in this study addresses the real-time and explainability challenges of mixed-integer optimization for asymmetric cooling source systems in super high-rise buildings. It forms a complete set of energy-saving control technologies that are "minute-level, implementable, and scalable," providing a direct engineering paradigm for central air-conditioning cooling source systems in similar super high-rise buildings.

Author Contributions: Conceptualization, J.Y.; methodology, J.Y.; software, J.Y. and A.X.; data curation, D.Z.; investigation, A.X. and D.Z.; resources, L.G.; writing—review and editing, J.Y. and D.Z.; validation, D.Z.; formal analysis, D.Z.; writing—original draft preparation, D.Z.; visualization, D.Z.; supervision, A.X.; project administration, J.Y.; funding acquisition, L.G. All authors have read and agreed to the published version of the manuscript.

Funding: This research was financially supported by the University-Level Scientific Research Fund Project of Guangzhou Maritime University (Grant Nos. K42022108, K42024047).

Data Availability Statement: The data presented in this study are available on request from the corresponding author.

Conflicts of Interest: Author Aiqin Xu is employed by Nanjing Fiberglass Research & Design Institute Co., Ltd. Author Lingjun Guan is employed by the Guangzhou Shijie Energy-Saving Technology Co., Ltd. The remaining authors declare that the research was conducted in the absence of any commercial or financial relationships that could be construed as a potential conflict of interest.

Abbreviations

The following abbreviations are used in this manuscript:

ANN	Artificial Neural Network
BA	Building Automation
BQ	Bivariate Quadratic Model
COP	Coefficient of Performance
CTI	Cooling Technology Institute
CSA	Cuckoo Search Algorithm
DE	Differential Evolution
DeST	Designer's Simulation Toolkit
DQN	Deep Q-Networks
EER	Energy Efficiency Ratio
GBR	Gradient Boosting Regressor
GA	Genetic Algorithm
HVAC	Heating, Ventilation, and Air Conditioning
HVAC&R	Heating, Ventilation, Air Conditioning, and Refrigeration
IPLV	Integrated Part-Load Value
MAE	Mean Absolute Error
MAPE	Mean Absolute Percentage Error
MLP	Multilayer Perceptron
MP	Multivariate Polynomial Regression

PLR	Part-Load Ratio
PSO	Particle Swarm Optimization
R^2	Coefficient of Determination
RMSE	Root Mean Square Error
SSA	Salp Swarm Algorithm
WOA	Whale Optimization Algorithm

References

1. Berardi, U. A cross-country comparison of the building energy consumptions and their trends. *Resour. Conserv. Recycl.* **2017**, *123*, 230–241. [CrossRef]
2. González-Torres, M.; Pérez-Lombard, L.; Coronel, J.F.; Maestre, I.R.; Yan, D. A review on buildings energy information: Trends, end-uses, fuels and drivers. *Energy Rep.* **2022**, *8*, 626–637. [CrossRef]
3. Chan, K.C.; Wong, V.T.; Yow, A.K.; Yuen, P.L.; Chao, C.Y. Development and performance evaluation of a chiller plant predictive operational control strategy by artificial intelligence. *Energy Build.* **2022**, *262*, 112017. [CrossRef]
4. Cha, J.H.; Park, J.; Yeon, S.H.; Yoon, Y.; Lee, K.H. Particle swarm optimization for multi-chiller system: Capacity configuration and load distribution. *J. Build. Eng.* **2024**, *98*, 110953. [CrossRef]
5. Parouha, R.P.; Verma, P. A systematic overview of developments in differential evolution and particle swarm optimization with their advanced suggestions. *Appl. Intell.* **2022**, *52*, 10448–10492. [CrossRef]
6. Tian, H.; Feng, M.; Fan, H.; Cao, R.; Gao, Q. Optimization control strategy for a central air conditioning system based on AFUCB-DQN. *Processes* **2023**, *11*, 2068. [CrossRef]
7. Xue, Z.; Yu, J.; Zhao, A.; Zong, Y.; Yang, S.; Wang, M. Optimal chiller loading by improved sparrow search algorithm for saving energy consumption. *J. Build. Eng.* **2023**, *67*, 105980. [CrossRef]
8. Hu, Y.; Qin, L.; Li, S.; Li, X.; Li, Y.; Sheng, W. Optimal chiller loading based on flower pollination algorithm for energy saving. *J. Build. Eng.* **2024**, *93*, 109884. [CrossRef]
9. Mirjalili, S.; Lewis, A. The whale optimization algorithm. *Adv. Eng. Softw.* **2016**, *95*, 51–67. [CrossRef]
10. Wang, L.; Li, J.; Yang, B. Low-carbon economic operation of integrated energy system considering carbon emissions of electric energy storage devices. *Sci. Technol. Eng.* **2021**, *21*, 2334–2342.
11. Wang, J.; Yue, Y.; Li, B. Microgrid optimization based on Archimedean chaotic elite whale algorithm. *Sci. Technol. Eng.* **2023**, *23*, 12577–12584.
12. Wang, J.; Xu, M. Bi-level planning of distributed power sources based on improved whale algorithm. *Sci. Technol. Eng.* **2024**, *24*, 11294–11302.
13. Gordon, J.M.; Ng, K.C. A general thermodynamic model for absorption chillers: Theory and experiment. *Heat Recovery Syst. CHP* **1995**, *15*, 73–83. [CrossRef]
14. Lee, T.S.; Lu, W.C. An evaluation of empirically-based models for predicting energy performance of vapor-compression water chillers. *Appl. Energy* **2010**, *87*, 3486–3493. [CrossRef]
15. Yik, F.; Lam, V. Chiller models for plant design studies. *Build. Serv. Eng. Res. Technol.* **1998**, *19*, 233–241. [CrossRef]
16. Reddy, T.A.; Andersen, K.K. An evaluation of classical steady state offline linear parameter estimation methods applied to chiller performance data. *HVACR Res.* **2002**, *8*, 101–124. [CrossRef]
17. Braun, J.E. Methodologies for the Design and Control of Central Cooling Plants. Ph.D. Thesis, University of Wisconsin-Madison, Madison, WI, USA, 1988. Available online: https://minds.wisconsin.edu/bitstream/handle/1793/46694/Braun1988.pdf (accessed on 1 January 2020).
18. Yan, D.; Zhou, X.; An, J.; Kang, X.; Bu, F.; Chen, Y.; Pan, Y.; Gao, Y.; Zhang, Q.; Zhou, H.; et al. DeST 3.0: A new-generation building performance simulation platform. *Build. Simul.* **2022**, *15*, 1849–1868. [CrossRef]
19. *GB 50189-2015*; Design Standard for Energy Efficiency of Public Buildings. China Architecture & Building Press: Beijing, China, 2015.

buildings

MDPI

Review

An Analytical Review of Humidity-Regulating Materials: Performance Optimization and Applications in Hot and Humid Regions

Dongliang Zhang [1], Tingyu Wang [1], Bo Zhou [2,*], Pengfei Zhang [2] and Jiankun Yang [3,*]

1 School of Ocean Engineering, Guangzhou Maritime University, Guangzhou 510725, China; zhangdongliang@gzmtu.edu.cn (D.Z.); wangtingyu@gzmtu.edu.cn (T.W.)
2 School of Environmental Science and Engineering, Suzhou University of Science and Technology, Suzhou 215009, China; zpf1911@163.com
3 School of Future Transportation, Guangzhou Maritime University, Guangzhou 510725, China
* Correspondence: zhoubo@mail.usts.edu.cn (B.Z.); yangjiankun@gzmtu.edu.cn (J.Y.)

Abstract

Humidity-regulating materials (HRMs) represent a promising class of passive, energy-efficient materials capable of autonomously modulating indoor environmental conditions, particularly in hot and humid regions where conventional HVAC systems account for up to 50% of building energy consumption. While prior reviews have focused on material classification and performance metrics, a systematic synthesis of performance optimization strategies and quantitative application outcomes remains lacking. This review addresses this gap by consolidating advances in HRM enhancement through material compounding, physical modification, and chemical functionalization, resulting in performance improvements such as a 70% increase in moisture absorption with 3% fiber addition, a 1.2-fold enhancement in adsorption capacity via pore optimization, and up to 50% energy savings in building applications. Furthermore, the integration of HRMs into radiant cooling systems elevates the dew point temperature difference by 181%, effectively mitigating condensation risks. Simulation tools—ranging from 1D to 3D multiphysics models—have advanced predictive accuracy for coupled heat and moisture transfer, supporting optimized material design and system integration. By systematically summarizing performance metrics, enhancement mechanisms, and real-world applications, this work provides a quantitative and structured reference for the development and deployment of next-generation HRMs in sustainable building systems.

Keywords: humidity-regulating materials; performance optimization; hot and humid regions; moisture buffering value; energy efficiency; pore structure; multiphysics modeling; phase change materials

1. Introduction

Building operations constitute a significant portion of global energy demand and carbon dioxide emissions, with figures reaching as high as 39% of total energy consumption and 40% to 48% of energy-related emissions in developed economies such as the United States [1,2]. Heating, ventilation, and air conditioning (HVAC) systems are among the most energy-intensive components in buildings, responsible for up to 50% of their total energy use [3]. The general method for dehumidification in conventional air conditioning systems requires cooling the air to its dew point to remove latent heat, after which it must

 https://doi.org/10.3390/buildings15234376

be reheated to the required supply temperature. This process of simultaneous cooling and reheating results in considerable energy inefficiency [4].

Humidity-regulating materials (HRMs) are an emerging class of smart materials that regulate ambient humidity autonomously and passively using their inherent physicochemical properties, requiring no external energy or mechanical systems [5]. This passive operation ensures zero energy consumption, making it consistent with green and sustainable development principles [6,7]. HRMs are mainly divided by composition and humidity regulation mechanism into inorganic mineral-based (e.g., silica gel, zeolite), inorganic salt-based (e.g., calcium chloride), organic polymer-based (e.g., cellulose), and composite types [8–11]. Inorganic mineral-based materials operate by physical adsorption, featuring fast moisture exchange but limited capacity. Inorganic salt-based materials achieve high capacity through chemical adsorption but often have poor desorption. Organic polymer-based materials provide high moisture storage but exhibit slow desorption. Composite humidity control materials combine the benefits of each, resulting in significantly enhanced performance [12].

Evaluation methods for the moisture absorption and release performance of HRMs are generally classified into three levels [13]: material, system, and room. At the material level, performance is evaluated based on fundamental hygrothermal properties, including the maximum equilibrium moisture content, water vapor diffusion coefficient, and sorption isotherms. At the system level, the dynamic moisture absorption and desorption capabilities of HRMs are investigated under typical boundary conditions involving variations in temperature and humidity. Among the metrics employed, the Moisture Buffering Value (MBV) is widely adopted as a key indicator. Literature reviews suggest that a higher MBV corresponds to improved humidity regulation performance and greater energy-saving potential. Although existing review studies [13–15] have systematically summarized the classification, mechanisms, and evaluation metrics of HRMs, they mainly focus on the fundamental properties of single-type materials and exhibit certain limitations. On the one hand, current reviews mainly address the performance of individual HRM categories, with insufficient exploration of optimization strategies, synergistic effects in composite materials, and the associated mechanisms and modeling approaches. On the other hand, there is a lack of systematic synthesis regarding the performance of optimized HRMs in practical applications. To address these research gaps, this paper systematically examines two key aspects: performance optimization methods for HRMs and the application performance of optimized HRMs across various fields.

The general purpose of this review is to systematically synthesize performance optimization strategies and application outcomes of humidity-regulating materials, bridging material science with building engineering practices. While comprehensive, this study has limitations: the rapidly evolving nature of advanced materials like MOFs may outpace this synthesis, and variations in testing conditions across cited studies limit direct performance comparisons.

2. Materials and Methods

This review employs a systematic and analytical approach to survey the scientific literature on humidity-regulating materials (HRMs). The primary literature search is conducted using major academic databases, including Web of Science, Scopus, and Google Scholar, with a focus on peer-reviewed journal articles, conference proceedings, and key books published predominantly between 2010 and 2025. Keywords such as "humidity-regulating materials", "moisture buffering value", "hygrothermal performance", "phase change materials (PCM)", and "building energy efficiency" are utilized to identify relevant publications.

The selection of literature for in-depth analysis is guided by the following criteria: (1) the study presented original research on the synthesis, modification, or performance characterization of HRMs; (2) the work provided quantitative data on key performance metrics, such as Moisture Buffering Value (MBV), equilibrium moisture content, or pore structure characteristics; or (3) the research demonstrated innovative applications of HRMs in building envelopes, HVAC systems, or related fields. The methodological framework of this review was structured around two core analytical themes: performance optimization strategies (categorized into material compounding, physical modification, and chemical enhancement) and application performance evaluation (encompassing building materials, radiant cooling terminals, and sensors), as illustrated in Figure 1. Furthermore, the principles of multiphysics modeling and simulation techniques for predicting coupled heat and moisture transfer were studied to establish a link between material properties and system-level performance.

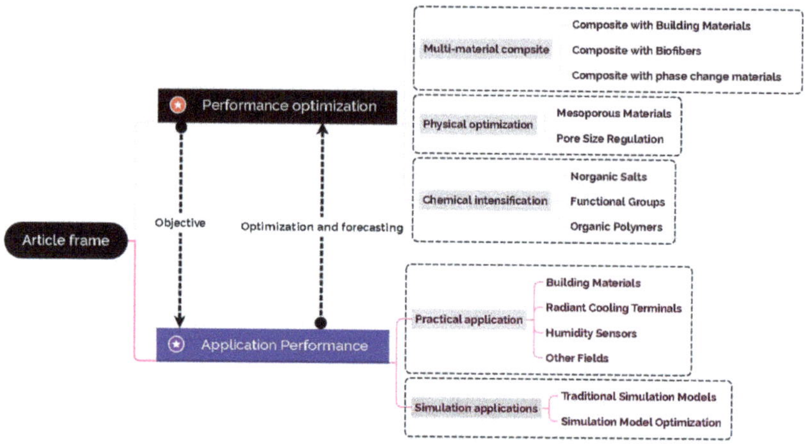

Figure 1. Article frame.

3. Performance Optimization of Humidity Control Materials

3.1. Multi-Material Composites

To improve the moisture absorption and desorption properties of materials, two or more humidity-regulating materials are frequently combined. Through physical mixing or chemical integration, complementary advantages emerged, leading to enhanced moisture regulation efficiency and moisture storage capacity. Such composite strategies not only elevate the practical value of materials but also offer technical support for their sustainable development and multi-scenario applications.

(1) Composite with Building Materials

The incorporation of humidity-regulating materials into conventional building envelope components represents a strategic approach to modifying the thermal and hygroscopic performance of existing structures. Furthermore, this composite strategy can enhance the mechanical strength of certain materials under thermal and moisture stresses.

Novais et al. [16] investigated the feasibility of utilizing cork as a lightweight aggregate in alkali-activated composites. Their findings indicated that a 75% cork content was critical for achieving peak performance in both moisture regulation and cost-effectiveness. Park et al. [17] found that incorporating biochar into mortar significantly improved the thermal and hygroscopic performance of porous materials. Their integrated experimental and simulation study established that a 4 wt% addition to mortar yielded the highest compressive strength while enhancing humidity regulation, regardless of the biochar type used. Meanwhile, Li et al. [18] investigated the water transport mechanism of lime-clay composites. They discovered that increasing lime content promoted a faster conversion of free water into bound and hydrated states, thereby improving moisture transfer.

However, since traditional building materials typically exhibit inherently limited humidity-regulating capacity, the resulting composites seldom exceed the maximum performance of any individual constituent. Consequently, this strategy predominantly serves to optimize existing structural systems, rather than substantially enhancing the intrinsic properties of the materials or significantly improving indoor thermal comfort.

(2) Composite with Biofibers

Biofibers, including hemp, straw, bamboo, and wood fibers, are derived from abundant agricultural byproducts. Owing to their renewable nature, low cost, high hygroscopicity, and minimal environmental impact, these materials have garnered extensive attention and found broad applications across various fields [19–22].

Bouferra et al. [23] improved the hygroscopic performance of clay-kaolin composites through the incorporation of plant fibers. Their experimental data indicated a positive correlation between moisture absorption capacity and fiber content: under conditions of 30 °C and 90% relative humidity (RH), a 3% fiber addition increased absorption by 70%, accompanied by an S-shaped isothermal adsorption curve. Wang et al. [24] fabricated inorganic humidity-regulating bricks by integrating poplar wood fibers with calcium hydroxide. The resulting composite demonstrated excellent moisture buffering capacity, which was attributed to an enhancement in capillary action during moisture transport. Gosselin et al. [25] investigated barley straw fibers and reported that low-density short fibers achieved a higher moisture buffer value (MBV) of 2.95 $g/(m^2 \cdot \Delta\%RH)$ compared to high-density long fibers. In addition, the performance of bio-based insulation materials and clayey plasters with olive fibers were studied as shown in Table 1 [26,27]. Bahammou et al. [28] found that biofiber-reinforced clay composites exhibited increased capillary water absorption at higher RH levels but reduced absorption at elevated temperatures. Biofiber composites are simple to prepare, eco-friendly, and cost-effective due to the reuse of agricultural waste. However, their durability may be inferior to other composites.

(3) Composite with phase change materials (PCM)

Phase change humidity control materials (PFCHRMs) represent a synergistic combination of humidity-regulating substances and phase change materials (PCMs), providing dual functionality in temperature and humidity regulation. Li et al. [29] developed a novel PFCHRM using a PCM for temperature control and hygroscopic MIL-100(Fe) for humidity regulation. This composite demonstrated exceptional reliability over repeated thermal-moisture cycles, highlighting its significant potential for applications in indoor environmental control and energy-efficient building systems.

A common challenge in PCM composites is leakage during phase transitions. And microencapsulation technology has been widely adopted to address this issue, thereby significantly enhancing the composite's structural stability [30–32]. For instance, Chen et al. [33] synthesized PCM microcapsules through a sol–gel method with methyltriethoxysilane as the precursor. These microcapsules were then integrated with a matrix of sepiolite, palygorskite, and zeolite to fabricate a phase change humidity control material (PFCHRM) that demonstrated robust thermal performance and effective moisture buffering capacity. In a related study, Wang et al. [34] developed a PFCHRM by incorporating microencapsulated capric acid PCM into a sepiolite-zeolite hygroscopic matrix. The resulting composite was characterized by a phase transition temperature of 31–32 °C, a high heat enthalpy of 123.91 J/g, and optimal humidity regulation (6.28% moisture absorption) when the sepiolite-to-zeolite mass ratio was 9:1. Fraisse et al. [35] incorporated a phase change humidity control material (PFCHRM)—comprising microencapsulated

PCMs and diatomite—into sintered hollow bricks and evaluated its performance against expanded polystyrene (EPS). Simulations showed PFCHRMs reduced internal temperature fluctuations, relative humidity, and total heat flux by 50%, highlighting their energy-saving potential. He et al. [36] formulated composite materials by incorporating PCM microcapsules into various matrices, including gypsum, zeolite, expanded vermiculite, and shell powder. Among these, the zeolite-based composite demonstrated a phase transition temperature range of 20–25 °C, which is ideal for room temperature and superior humidity control. By providing synergistic control over both temperature and humidity, PFCHRMs exhibit a distinct advantage over traditional single-function humidity-regulating materials, underscoring their significant potential for advanced applications in building indoor environment creation.

Table 1 summarizes the moisture-related parameters of optimized humidity-regulating materials. Among these parameters, Moisture Buffering Value (MBV) serves as a key performance indicator, where a higher value denotes a greater capacity for humidity regulation. The equilibrium moisture content reflects the moisture capacity of composite materials. Generally, the equilibrium moisture content of materials increases with the elevation of ambient relative humidity.

Table 1. Performance parameters of composite humidity-regulating materials.

Material	Reinforcement Material	MBV (g/(m^2·%RH))	Equilibrium Moisture Content	Relative Humidity (RH)	References
Alkali-activated	Cork	1.89	/	/	[16]
Mortar	Biochar	0.7	40.29 kg/m^3	80%	[17,18]
Mortar	Corn Cob/ Wood Wool	1.116/ 0.663	35%/15%	90%	[26]
Clay	Olive Fiber	1.739	30 kg/m^3	80%	[27]
Inorganic humidity-regulating Brick	Poplar Wood Fiber	1.917	5.5%	80%	[24]
PCM	MIL-100(Fe)	10.2	29.8%	60%	[29]
PCM Microcapsules	Vesuvianite/Sepiolite/ Zeolite	1.145/0.78/ 0.514	/	/	[34]
PCM Microcapsules	Sepiolite-Zeolite Powder	/	6.28%	99%	[34]
PCM Microcapsules	Gypsum/Zeolite/ Expanded Vermiculite/Shell Powder	0.495/4.36/ 2.82/0.433	1%/10%/12%/1%	80%	[36]

Notes: MBV: Moisture Buffer Value (g/(m^2·%RH)). Equilibrium Moisture Content: Expressed as mass percentage (%) or volumetric content (kg/m^3). Relative Humidity: Test conditions for equilibrium moisture measurement. "/": Data not provided or not applicable.

3.2. Physical Optimization

The pore structure of humidity-regulating materials is categorized into one-end open pores and two-end open pores. Since one-end open pores lack a hysteresis loop (due to overlapping adsorption and desorption curves), which deviates from the ideal isothermal adsorption characteristics of humidity-regulating materials, consequently, two-end open pore structures are generally preferred. In such materials, capillary condensation occurs at varying temperatures, with a specific correlation between pore radius and relative humidity (RH). Dr. Yang Qian and her team at Southwest University published a study in Nature [37], demonstrating that the Kelvin equation remains valid even in sub-nanometer channels. A systematic analysis of IUPAC-classified adsorption isotherms indicates that materials exhibiting Type IV or V behavior exhibit a pronounced affinity for humidity regulation. The study further quantified that, within the 40–70% RH range, the critical pore diameters for capillary condensation and desorption in two-end open pores fall within 1.16–2.96 nm and 2.32–5.92 nm, respectively. By accounting for the presence of pre-adsorbed water molecular layers during dynamic moisture exchange, the optimal pore size for effective humidity regulation can be inferred to lie between 3.0 and 7.4 nm.

(1) Mesoporous Materials

Mesoporous materials, characterized by pore diameters of 2–50 nm, inherently fall within the optimal size range for humidity regulation, making them widely applicable in this field. The most well-established synthesis routes, such as sol–gel and hydrothermal methods, are frequently combined with microwave-assisted or sonochemical techniques to precisely control and enhance their structural and functional properties.

Zhang et al. [38] prepared mesoporous materials via a one-pot hydrothermal method for humidity sensing applications, systematically identifying the optimal precursor ratios to enhance sensing performance. In a separate study, Lu et al. [39] synthesized low-cost mesoporous silica nanoparticles (MSNs) from waste materials using hydrothermal methods, achieving an average pore size of 3.98 nm with uniform porosity.

Disordered mesoporous materials suffer from irregular pore structures and broad size distributions, leading to unstable adsorption performance. To address this, researchers have either optimized disordered mesopores or directly synthesized ordered mesoporous materials with highly aligned, monodisperse pores. For example, Gao et al. [40] developed crystalline mesoporous magnesium silicate via a sol–gel route, using carbon coating to prevent pore collapse during high-temperature crystallization.

(2) Pore Size Regulation

To enhance performance, pore sizes of other humidity-regulating materials are often tailored to the optimal range. It is demonstrated that foaming treatment optimizes the pore structure of humidity-regulating materials, specifically by increasing porosity and improving pore size distribution. This process led to the formation of more micropores (2–4 nm in diameter) and a reduction in mesopores ($\geq$15 nm in diameter). Such structural modifications significantly improved the material's hydrothermal performance, increasing its maximum adsorption capacity by 1.2 times and water vapor permeability by 1.09 times. Meanwhile, Zhou et al. [41] developed a hierarchically porous composite aerogel via a freeze-drying method. Experimental results demonstrated that this material effectively maintained indoor humidity within an optimal range of 50–65%. Moreover, even after 30 cyclic humidity tests conducted between 20% and 90% relative humidity, its moisture regulation performance remained stable.

Metal–organic frameworks (MOFs) are an emerging porous materials distinguished by their exceptional pore size distribution and structural tunability. Qin et al. [42] developed a precision humidity control material (PHRM) based on an MOF structure. Characterization results indicated that the synthesized MOF material possesses an average pore size of approximately 2 nm, exhibiting an S-shaped isotherm, high porosity, and a remarkable water vapor uptake capacity of 1.62 g/g at 80% relative humidity. In a related study, Hou et al. [43] fabricated a localized humidity pump using MOFs, which demonstrated efficient moisture transfer from low-humidity to high-humidity spaces under environmental conditions of 22.8 °C and 60% RH. Ge et al. Meanwhile, Ge et al. [44] utilized the highly ordered mesoporous channels of MCM-41 to provide uniform nucleation sites for MOF growth, successfully synthesizing a nanocomposite with an ultrahigh specific surface area (3188.14 cm^2/g) through hydrothermal methods.

Pore-size preparation methods for the aforementioned mesoporous humidity-regulating materials are summarized in Table 2. Primary techniques include hydrothermal synthesis, template methods, sol-gel processes, foaming, and freeze-drying. Notably, the most probable pore size (the pore diameter with the highest occurrence frequency in porous materials) is consistently less than 10 nm for all methods, which falls within the optimal pore size range for achieving superior humidity regulation performance as discussed above.

Table 2. Pore size regulation methods for humidity-regulating materials.

Material	Method	Most Probable Pore Size	References
Inorganic Compounds	Template Method	3.93 nm	[38]
Silica	Hydrothermal Synthesis	8.4 nm	[39]
Magnesium Silicate	Sol–Gel Method	2.58 nm	[40]
Palygorskite/Wood Fiber Composite	Freeze-Drying	10 nm	[41]
Metal–Organic Framework	Hydrothermal Synthesis	2 nm	[43]
MIL-101/MCM-41 Composite	Template Method	3.2 nm	[45]

Notes: Most Probable Pore Size: Dominant pore size range for each material.

3.3. Chemical Enhancement

(1) Inorganic Salts

Among various humidity-regulating materials, natural inorganic porous materials stand out for their cost-effectiveness and durability, making them suitable for large-scale building applications. However, these materials generally suffer from limited moisture capacity. To enhance their hygroscopic performance, the most common approach involves impregnation with highly hygroscopic inorganic salts.

Liu et al. [45] developed smart humidity-regulating wall tiles using sepiolite with $CaCl_2$ as an additive. Experimental tests under real-world conditions demonstrated that these white wall tiles successfully reduced daily indoor humidity fluctuations by over 20% in complex outdoor environments. In a related study, Yang et al. [46] investigated the dehumidification performance of LiCl solutions and examined their effects on the basic properties (e.g., thermal conductivity), hygroscopicity (adsorption/desorption), and moisture buffering performance of gypsum-based materials. The results indicated that a small addition of LiCl (1–8%) significantly improved the fundamental and hygroscopic properties of gypsum boards while avoiding noticeable microstructural alterations.

(2) Functional Groups

Grafting reaction refers to a type of copolymerization reaction that produces graft copolymers. Graft copolymers consist of two different polymer chains connected by chemical bonds. By introducing hydrophilic functional groups such as hydroxyl (-OH), carboxyl (-COOH), and amino (-NH_2) onto mesoporous materials through grafting, their underlying support structures can be modified. This grafting approach enhances the material's moisture—absorption capacity by forming hydrogen bonds between the functional groups and water molecules. Additionally, the introduction of appropriately sized organic functional groups enables the adjustment of pore sizes in mesoporous materials.

Ming-Jui Hung et al. [47] conducted amine functionalization using varying concentrations of (3-aminopropyl)triethoxysilane under a nitrogen atmosphere at 120 °C. The study found that Al—MCM—41 molecular sieves with 5.0% grafted amine groups exhibited excellent structural properties. These properties included a high specific surface area of 870.23 m^2/g, a pore volume of 1.41 cm^3/g, and an average pore size of 4.74 nm. Furthermore, these molecular sieves demonstrated superior pore uniformity and an outstanding equilibrium moisture content, reaching up to 39.4 kg/kg.

(3) Organic Polymers

Polymers such as polyacrylic acid (PAA), polyacrylamide (PAM), and carboxymethyl cellulose (CMC) leverage hydrophilic groups and 3D networks to achieve excellent moisture regulation performance. These materials can be flexibly processed through various methods, including blending, copolymerization, and lamination. For instance, Xue et al. [48]

modified vermiculite with acrylic acid/acrylamide via inverse suspension polymerization, creating a novel humidity-regulating composite. Yue et al. [49] synthesized lightweight, tough building materials using expanded perlite (EP) and CMC, where CMC's crosslinking network and osmotic pressure boosted self-regulating performance. Synergistic effects from $CaCl_2$ and CMC further enhanced moisture uptake.

4. Application Performance

4.1. Practical Applications

Since the 1990s, HRMs have been integrated into the interior surfaces of building envelopes. In conjunction with conventional enclosure assemblies, these moisture-responsive layers supplement active humidification systems, thereby improving indoor thermal comfort while simultaneously reducing the energy demand of HVAC equipment. Up to now, the applications of these materials have expanded to fields such as construction, air conditioning systems, sensing, and wastewater treatment.

(1) Building Materials

The integration of HRMs into building walls aims to effectively control indoor humidity, enhancing occupant comfort and extending the structural lifespan of buildings. By adding these materials to construction components, they can absorb and release moisture from the air, thereby stabilizing indoor humidity levels. As depicted in Figure 2, humidity-regulating wall has been applied in museums.

Figure 2. Humidity-regulating wall in museum.

For example, Liu et al. [50] developed intelligent humidity-regulating wall tiles using sepiolite with $CaCl_2$ as an additive. The low-temperature sintering process created a highly hygroscopic internal structure within the tiles. Tests conducted in a real-world model house demonstrated that these tiles could reduce daily indoor humidity fluctuations by over 20%. The team also modified the tiles with silane to create a superhydrophobic surface, enhancing their self-cleaning properties while effectively preventing leakage of the $CaCl_2$ solution. Another researcher Fraine et al. [35] prepared a phase-change humidity control material by combining microencapsulated phase-change materials (MPCM) with diatomite, and then incorporated into sintered hollow bricks. This study indicated that compared to traditional expanded polystyrene, the MPCM/diatomite composite material exhibited energy-saving potential of up to 50%.

Additionally, Zhou et al. [51] employed a sol–gel method to composite zeolite with polymer materials, producing a hygroscopic and thermal insulation material with a porous surface and uniform, single-sized pore distribution. And when applied to prefabricated houses, the study found that during winter, this composite material effectively mitigated the impact of outdoor temperature and humidity fluctuations on indoor conditions, stabilizing air parameters such as temperature, relative humidity, and moisture content.

Furthermore, the long-term durability HRMs in building envelopes are critically dependent on their waterproofness and resistance to alternating moisture and drying cycles. Prolonged exposure to high-humidity environment and recurrent wetting-drying stresses can lead to material performance degradation. For instance, Sara C. [52] experimentally and numerically demonstrated that a compressed raw earth block wall effectively provides passive regulation of indoor temperature and humidity while also functioning as a moisture barrier. Hongqiang et al. [53] conducted an experimental investigation on the waterproof performance of building reed-based insulation materials prepared by carbonized reed, alkalic activator, and self-developed waterproof agent. In addition, the applications of Metal–Organic Frameworks have been extended to cultural heritage preservation, encompassing environmental control through humidity regulation, adsorption of harmful gases and particulate matter, as well as surface waterproofing and antibacterial protection [54].

(2) Radiant Cooling Terminals

Radiant cooling systems have gained wide attention due to their energy efficiency and humane thermal comfort benefits. This kind of air conditioning system utilizes chilled water at moderate temperatures to generate uniform thermal distribution while minimizing air movement, reducing noise, and eliminating drafts. However, condensation will occur when indoor humidity fluctuates rapidly or surface temperatures drop too low, posing a major obstacle to the widespread application of radiant cooling systems.

Such as, Chen et al. [55] addressed the inability of conventional radiant terminals to handle latent heat loads by integrating humidity-regulating materials into the system as shown in Figure 3, proposing a novel hygrothermal radiant terminal with moisture storage capacity to mitigate condensation risks. Zhang et al. [56] investigated the condensation-suppression performance of an MOF-based composite hygroscopic coating applied to surfaces in a typical radiant cooling room. It was found that compared to the uncoated ceiling, the coated material with a higher Moisture Buffering Value (MBV) could significantly increase the average difference between surface temperature and dew point temperature—by up to 181%. Additionally, compared to a structure without desiccants, the average temperature difference between the surface and dew point increased by 2.11 °C. In addition, sepiolite based humidity-control material and sepiolite based humidity-control coating have been prepared for radiant cooling panel to alleviate its condensation problem [55,57].

Figure 3. Humidity-regulating radiant ceiling.

The humidity-regulating coating, composed of advanced moisture-adaptive materials, can significantly improve a building's energy utilization efficiency. This innovation overcomes the condensation default of traditional radiant cooling systems and substantially reduces system energy consumption.

(3) Humidity Sensors

The integration of HRMs has been successfully employed to enhance the accuracy and stability of humidity sensors. Humidity sensors—devices that monitor atmospheric moisture levels—typically rely on resistive or capacitive sensing mechanisms. The incorporation of HRMs effectively mitigates the sensor's temperature sensitivity, leading to significant improvements in measurement precision and response speed.

Thus, Zhen et al. [58] synthesized a tin oxide/diatomite composite material via a hydrothermal method and investigated its humidity-sensing and control properties. It was found that the honeycomb structure of diatomite provided the composite with a large specific surface area for H_2O molecule adsorption while enabling charge transfer between embedded tin oxide nanoparticles. This design significantly enhanced the sensor's humidity sensitivity and response time. In addition, MOF-based humidity sensors have been studied and developed for humidity monitoring in indoor environments or industrial fields in recent years [59–62]. An enhanced-performance relative humidity sensor using MOF-801 photonic crystals was developed by Kuo Z. et al. [59]. Wu et al. [60] conducted an in-depth study on six MOF-based humidity sensors, identifying their shared characteristics, including strong hydrophilicity, high porosity, and exceptional stability. Three free-base Zr porphyrin metal–organic frameworks (MOF-525, MOF-545, and NU-902) were studied by Nicholaus for their efficacy in QCM-based relative humidity sensing [61]. Ke et al. [62] addressed the demand for highly sensitive low-humidity sensors by directionally constructing long-range ordered ionic liquid (IL) functionalized MOF-303, which enabled a sensor with exceptional performance, including a response of 3075, and its mechanism was elucidated through characterization and CIS analysis.

(4) Other Fields

Most HRMs possess hierarchical porosity with high specific surface area, endowing them with exceptional adsorption capacity. When applied to wastewater treatment containing abundant organic compounds and microorganisms, these materials can effectively adsorb and decompose organic pollutants.

For instance, Alsuhaibani et al. [63] developed a composite sponge by encapsulating sulfadoxine in chitosan-modified MOF material. This sponge had a large surface area of 862.87 m^2/g and showed effective performance as an adsorbent for removing contaminant ions in wastewater systems. Zhao et al. [64] synthesized a biochar-based composite material. Their research showed that under optimal conditions (impregnation ratio of 1:2, dosage of 50 mg, and initial solution pH of 4.0), this composite achieved a remarkable pollutant removal efficiency of 97.56%.

HRMs have been demonstrated to significantly enhance the performance of humidity sensors. However, the current literature on this subject remains limited, necessitating further in-depth research.

4.2. Simulation Applications

Practical performance of HRMs not only depends on their intrinsic properties but also crucially on their heat and moisture transfer behavior during application. Given the long testing cycles and high costs associated with experimental setups, numerical simulation has emerged as an effective alternative research approach.

By simulation applications, researchers can analyze extensive testing data from real-world material deployment scenarios to uncover the underlying physical and chemical mechanisms responsible for performance limitations. These mechanisms are then mathematically modeled, solved using appropriate algorithms, and iteratively refined to accurately predict material performance in practical applications. Consequently, in simulation studies of humidity-regulating materials, the validity of moisture transfer models directly determines the reliability of performance predictions.

(1) Traditional Simulation Models

(1) Moisture Transfer Models

Fick's law states that the diffusion flux through a unit cross-sectional area perpendicular to the diffusion direction is proportional to the concentration gradient at that cross-section, with higher gradients yielding greater fluxes. Lewis [65] proposed that surface evaporation reduces internal moisture content and alters temperature, making diffusion coefficients variable. Stefan [66] extended Fick's theory into Stefan's law, assuming isotropic porous materials with constant physical properties and vapor diffusion driven by partial pressure, while neglecting liquid water transport. Liquid water in porous media primarily moves via surface diffusion and capillary forces, with Darcy's law describing its transport. Buckingham [67] introduced the concept of unsaturated capillary flow driven by capillary potential, and Miller [68] linked capillary transport to surface tension and viscosity, both pressure-dependent functions. Most humidity-regulating materials are porous, with liquid water transport driven by capillary potential, diffusion, and hydrostatic pressure. Capillary models excel at describing moisture transfer under low humidity but become less accurate under high humidity, where diffusion and permeation dominate.

In common commercial simulation software (as listed in Table 3), EnergyPlus, MATCH, and hygIRC focus on one-dimensional heat and mass transfer. They are suitable for simplified models (e.g., gradient analysis along wall thickness) or specific scenarios (e.g., energy consumption simulation, hygroscopic material calculations). MATCH ignores thermal and moisture convection effects, which makes it ideal for foundational theoretical research. Meanwhile, hygIRC emphasizes thermal–air–moisture coupling and is suitable for assessing building durability.

Table 3. Commercial Software for Heat and Moisture Transfer in Hygroscopic Materials.

Software Name	Transfer Dimensions	Core Features and Advantages	References
WUFI	1D/2D/3D	Engineering-grade coupled heat-moisture model; mature for wall system analysis.	[69–71]
COMSOL	1D/2D/3D	Multi-physics coupling (heat, moisture, radiation, convection); highly customizable.	[72,73]
EnergyPlus	1D	Whole-building energy simulation; simplifies wall heat-moisture transfer to 1D.	[74,75]
MATCH	1D	Transient hygrothermal calculations for hygroscopic materials; combines finite volume method with Fick's law.	[76]
TRNSYS	1D/2D	Modular dynamic simulation; supports system integration and control strategy optimization.	[73]
hygIRC	1D	Fully coupled heat-air-moisture (HAM) model; emphasizes building durability.	[77]
DELPHIN	1D/2D	Fully coupled heat-air-moisture (HAM) model; emphasizes moisture impact on building durability.	[69,70,78]

Notes: Transfer Dimensions: 1D (one-dimensional), 2D (two-dimensional), 3D (three-dimensional).

WUFI-PLUS, COMSOL, TRNSYS, and DELPHIN support 1D to 3D transfer, making them applicable for complex structures (e.g., building envelopes, porous media) involving multiphysics coupling simulations. COMSOL is notable for its multiphysics coupling capabilities, allowing for the simultaneous analysis of heat conduction, convection, radiation, and moisture diffusion. DELPHIN, conversely, prioritizes the combined transfer of liquid water and water vapor and their impact on heat transfer.

(2) Coupled Heat-Moisture Models

Coupled heat-moisture transfer models account for interactions between humidity, temperature, and other factors, providing a holistic description of heat and moisture transport. In porous media, heat and moisture transfer are interdependent, necessitating coupled models for realistic simulations. Luikov's theory [79] integrates total pressure gradients, temperature gradients, concentration gradients, molecular transport, and capillary effects into a coupled model. However, its idealized nature requires simplification for practical use. Philip and De Vries [76,80] established a mathematical model driven by temperature and moisture gradients. In this model, liquid water moves via capillary and adsorption forces, and vapor diffuses through pores. Liu and Chen [81] developed a model using relative humidity as the driving potential, treating liquid water flow as a capillary pressure-driven "stream" governed by Darcy's law.

(2) Simulation Model Optimization

The accuracy and reliability of moisture transfer models depend on their validity. Optimizing these models for specific materials enhances the precision of simulations by improving the fidelity of the mathematical and physical models.

(1) Mathematical Models

To enhance the accuracy of numerical simulations, researchers have innovatively developed new models and methods for studying moisture transfer and humidity control processes. Currently, the optimization of moisture transfer models in porous media primarily focuses on multiphysics coupling, heterogeneity considerations, and mechanistic modeling.

Traditional theories of heat and moisture transfer in porous media are based on continuum assumptions, employing volume averaging that neglects the heterogeneity of microscopic pore structures and matrices. To address this limitation, Yuan et al. [82] adopted a discrete pore network model to simulate soil temperature and moisture distribution. This approach demonstrated superior accuracy compared to continuum-scale methods and better agreement with experimental data.

Additionally, Bianchi Janetti et al. [83] compared pore network models with continuum approaches. They revealed the microscopic mechanisms by which contact angle variations influence spontaneous imbibition in porous materials, further validating the advantages of discrete methods for interface-dominated problems. While continuum methods fail to capture pore-scale effects, discrete approaches (e.g., pore network models) can better characterize microstructural influences.

Fang et al. [84] developed a transient model addressing coupled heat, air, and moisture transfer through multilayer porous media. This model resolves non-steady-state moisture transfer characteristics under transient humidity shocks, providing theoretical support for defining material performance boundaries in dynamic applications. Choi et al. [85] quantified the impact of different driving forces on hygrothermal predictions, comparing indoor humidity variations under seasonal environmental changes. Their study confirmed that coupled Heat-Air-Moisture (HAM) models using water potential as the driving force more accurately simulate the hygrothermal behavior of building envelopes. Son Thai Pham et al. [86] established a DEM-based triangulated pore-scale model for drying particle aggregates, uncovering the influence of particle size distribution on capillary force evolution. This model constructs a quantitative relationship chain linking "particle size distribution → pore structure → capillary forces → moisture transfer performance," enabling precise material design based on microscopic mechanisms.

As depicted in Figure 4, in porous materials, moisture exists in both vapor and liquid phases at interfaces and within pores, with transfer typically occurring as a gas–liquid two-phase flow. Conventional models often treat these phases independently, despite their coupled behavior. Zhao et al. [87] proposed a mesoscale numerical method for coupled heat–moisture transfer in building envelopes, introducing a novel multiphase model that incorporates heat exchange, vapor migration, freezing, and evaporation/condensation effects. Sadeghi et al. [88] extended Lee's multiphase model into 3D to simulate absorption/desorption in porous substrates, demonstrating that increased porosity enhances water penetration rates. Such multiphase models are critical for predicting internal heat-moisture transfer in hygroscopic walls and optimizing material performance.

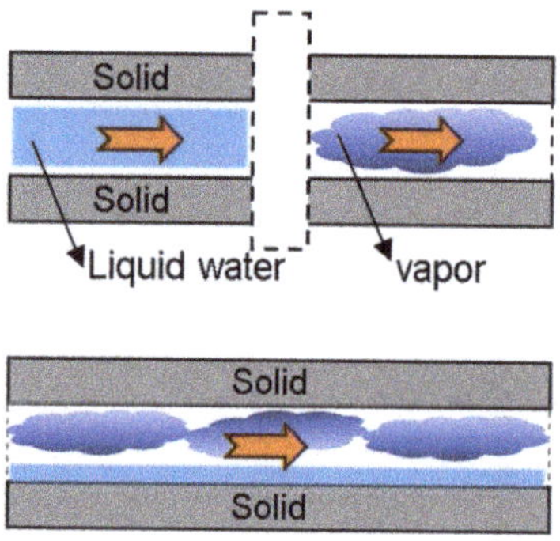

Figure 4. Forms of wet migration in pores.

The Kunzel model [89], which uses relative humidity as the driving potential, is a simplified approach widely cited for describing the behavior of hygroscopic materials, particularly under non-isothermal and non-isohumidity conditions in building materials [90–92].

Liu et al. [81] advanced this field with a coupled heat-moisture (CHM) model that employs vapor pressure and temperature as driving potentials. By avoiding approximations between air moisture content and relative humidity through total differential formulation and optimizing equation coefficients for solvability, this model effectively predicts the hygrothermal processes of walls. Some scholars refer to it as the "Liu model," noting that it avoids potential discontinuities at the interfaces between dissimilar porous media.

Comparative studies by Dong et al. [93] validated the accuracy of both models within the hygroscopic range. The Liu model shows superior precision, while the Kunzel model exhibits limitations near over-hygroscopic regions. However, Janssen [94] later argued that the two models are fundamentally similar.

(2) Physical Models

While the aforementioned simulation optimizations primarily refine classical moisture transfer models, another group of researchers has adopted innovative approaches by constructing realistic geometric structure models to significantly enhance the accuracy of numerical simulations.

Bennai et al. [95] employed X-ray tomography to reconstruct 3D porous material structures, enabling mesoscale-coupled heat-moisture transfer simulations via COMSOL Multiphysics. Similarly, Dahanni et al. [96] incorporated authentic 3D wood microstructures to simulate internal hygrothermal transfer, utilizing X-ray microtomography for physical model reconstruction. Galy et al. [97] numerically generated 3D structures with dispersed particles featuring point/surface contacts using the diffusion-limited cluster-cluster aggregation (DLCCA) method. The synthesized mesoporous structures demonstrated high fidelity to experimental characterizations in specific surface area, mean pore size, and size distribution. Muhammad Nasir et al. [98] conducted 3D drying experiments on wet porous media using micro-CT technology, revealing wettability effects during pore-scale drying processes.

This paradigm shift from idealized to true-structure modeling marks a critical evolution in hygrothermal simulation, particularly for applications that demand high precision in humidity-regulating material design and performance prediction. With the application of emerging intelligent technology in buildings [99–102], this field now confronts the challenge of developing smart simplification protocols to strike a balance between physical accuracy and computational practicality.

5. Discussion

This review provides a systematic synthesis of advances in humidity-regulating materials (HRMs). While prior review papers have cataloged the classification, mechanisms, and basic evaluation metrics of various HRMs [13–15], they have predominantly remained within the domain of material-level analysis. The scientific novelty of this work lies in its structured, two-tiered analytical framework that systematically bridges the gap between material science and engineering application. First, it moves beyond mere classification to consolidate and categorize performance optimization strategies into three coherent pathways: multi-material compounding, physical pore-structure regulation, and chemical functionalization, offering a clear roadmap for material enhancement. Second, and more critically, it establishes a direct link between these optimized material properties and their tangible performance in real-world systems, such as building envelopes, radiant cooling terminals, and sensors.

This systematization of knowledge yields significant practical implications for the engineering construction industry. By quantitatively summarizing performance gains—such as the 50% reduction in internal heat flux with PFCHRM-filled bricks [35], the 181% increase in dew-point temperature difference with MOF-based coatings [56], or the 20% reduction in indoor humidity fluctuations achieved by sepiolite-$CaCl_2$ tiles [50]—this review translates material science breakthroughs into measurable engineering outcomes. It compellingly argues for the role of HRMs not merely as passive components but as active contributors to energy conservation, operational stability, and durability of building systems. Furthermore, the critical analysis of multiphysics models and simulation tools provides engineers and architects with a practical guide for selecting appropriate predictive methods, thereby de-risking the integration of innovative HRMs into building designs. Ultimately, this work underscores a paradigm shift towards designing buildings as holistic, climate-responsive systems, where advanced materials are central to achieving sustainability and resilience goals, particularly in the challenging environments of hot and humid regions.

6. Conclusions

(1) Humidity-regulating materials demonstrate exceptional moisture control capabilities and show significant potential in building envelope applications. Physical enhancement methods mainly involve optimizing pore sizes to the range of 3.0–7.4 nanometers to achieve optimal humidity regulation. Mesoporous materials, due to their suitable pore characteristics, have been widely used in the preparation of humidity-regulating materials. Chemical enhancement approaches include inorganic salt impregnation and functional group grafting techniques, which aim to improve moisture absorption properties. Moreover, when organic polymer materials such as polyacrylic acid and carboxymethyl cellulose are combined with porous materials through composite techniques like blending and copolymerization, the moisture absorption performance can be further enhanced.

(2) Humidity-regulating materials play crucial roles in fields such as construction, radiative cooling systems, sensors, and water treatment. In building walls, they regulate indoor humidity to enhance comfort levels. In radiant cooling systems, they reduce the risk of condensation while improving energy efficiency. For humidity sensors, they enhance measurement accuracy and stability. In water treatment applications, they efficiently adsorb and decompose pollutants in wastewater, thus increasing treatment efficiency. With broad application prospects, future research may focus on the multifunctional integration of these materials to meet the high-efficiency demands across diverse fields.

(3) Significant advancements have been made in modeling heat and moisture transfer in humidity-regulating materials. Simulation approaches have evolved from foundational laws (Fick's, Darcy's) to sophisticated models that account for multiple coupled factors. To improve predictive accuracy, researchers are integrating more environ-

mental variables and material parameters, while refined geometric modeling enables precise simulation of moisture regulation. These developments provide a robust theoretical foundation for applying these materials in built environments.

(4) Future HRM research will shift from passive materials to intelligent, multi-functional systems responsive to temperature, light, and pollutants. Key directions include integrating air purification and energy harvesting, advancing phase-change materials, employing AI and cross-scale modeling for inverse design, and prioritizing circular economy principles using industrial and agricultural waste. In construction applications, HRMs are expected to be incorporated as standardized components in prefabricated and 3D-printed building systems. Key goals include establishing HRMs in passive building standards, developing anti-condensation solutions for radiant cooling, validating performance through large-scale demonstrations, and creating practical design guidelines for applications in hot and humid regions.

Author Contributions: Supervision, B.Z.; methodology, D.Z., T.W. and J.Y.; formal analysis, D.Z. and T.W.; investigation, D.Z. and P.Z.; resources, P.Z.; writing—original draft preparation, D.Z. and P.Z.; writing—review and editing, B.Z.; project administration, B.Z. and J.Y.; funding acquisition, D.Z. and J.Y. All authors have read and agreed to the published version of the manuscript.

Funding: This research was financially supported by the University-Level Scientific Research Fund Project of Guangzhou Maritime University (Grant Nos. K42024047) and the National Natural Science Foundation of China (Grant Nos. 52506074).

Data Availability Statement: The data presented in this study are available on request from the corresponding author. The data are not publicly available due to the nature of a review article. All data analyzed in this study were derived from the published literature cited in the reference list.

Conflicts of Interest: The authors declare no conflict of interest.

Abbreviations

The following abbreviations are used in this manuscript:

HVAC	Heating, ventilation, and air conditioning
RH	Relative humidity
HRM	Humidity-regulating material
MBV	Moisture buffering Value
PFCHRM	Phase change humidity control material
PCM	Phase change material
EPS	Expanded polystyrene
MSN	Mesoporous silica nanoparticle
MOF	Metal–organic frameworks
PAA	Polyacrylic acid
PAM	Polyacrylamide
CMC	Carboxymethyl cellulose
EP	Expanded perlite
CHM	Coupled heat-moisture
DLCCA	Diffusion-limited cluster-cluster aggregation

References

1. Mayernik, J.; Office of Energy Efficiency Renewable Energy. *Buildings Energy Data Book*; National Renewable Energy Laboratory, Ed.; National Renewable Energy Laboratory: Golden, CO, USA, 2015.
2. Shaikh, P.H.; Nor, N.B.M.; Nallagownden, P.; Elamvazuthi, I.; Ibrahim, T. A review on optimized control systems for building energy and comfort management of smart sustainable buildings. *Renew. Sustain. Energy Rev.* **2014**, *34*, 409–429. [CrossRef]
3. Chua, K.J.; Chou, S.K.; Yang, W.M.; Yan, J. Achieving better energy-efficient air conditioning—A review of technologies and strategies. *Appl. Energy* **2013**, *104*, 87–104. [CrossRef]

4. Tomson, A.E.; Sokolova, T.V.; Sosnovskaya, N.E.; Pekhtereva, V.S.; Goncharova, I.A.; Arashkova, A.A.; Kozhich, D.T. Prospects for the use of modified peat for indoor humidity control. *Solid Fuel Chem.* **2017**, *51*, 321–325. [CrossRef]
5. Pelliccia, G.; Baldinelli, G.; Bianconi, F.; Filippucci, M.; Fioravanti, M.; Goli, G.; Rotili, A.; Togni, M. Characterisation of wood hygromorphic panels for relative humidity passive control. *J. Build. Eng.* **2020**, *32*, 101829. [CrossRef]
6. Sultan, M.; El-Sharkawy, I.I.; Miyazaki, T.; Saha, B.B.; Koyama, S. An overview of solid desiccant dehumidification and air conditioning systems. *Renew. Sustain. Energy Rev.* **2015**, *46*, 16–29. [CrossRef]
7. Padfield, T. Humidity Buffering of the Indoor Climate by Absorbent Walls. In Proceedings of the 5th Symposium on Building Physics in the Nordic Countries, Gothenburg, Sweden, 24–26 August 1999.
8. Saatchi, D.; Oh, S.; Oh, I.-K. Biomimetic and Biophilic Design of Multifunctional Symbiotic Lichen-Schwarz Metamaterial. *Adv. Funct. Mater.* **2023**, *33*, 2214580. [CrossRef]
9. Vu, D.H.; Wang, K.-S.; Bac, B.H.; Nam, B.X. Humidity control materials prepared from diatomite and volcanic ash. *Constr. Build. Mater.* **2013**, *38*, 1066–1072. [CrossRef]
10. Young, J.F. Humidity control in the laboratory using salt solutions—A review. *J. Appl. Chem.* **1967**, *17*, 241–245. [CrossRef]
11. Wang, R.M.; Wang, J.F.; Wang, X.W.; He, Y.F.; Zhu, Y.F.; Jiang, M.L. Preparation of acrylate-based copolymer emulsion and its humidity controlling mechanism in interior wall coatings. *Prog. Org. Coat.* **2011**, *71*, 369–375. [CrossRef]
12. Hu, Z.; Zheng, S.; Tan, Y.; Jia, M. Preparation and characterization of diatomite/silica composite humidity control material by partial alkali dissolution. *Mater. Lett.* **2017**, *196*, 234–237. [CrossRef]
13. Cui, Y.; Zhang, Y. Review of evaluation methods for moisture absorption and desorption properties of humidity control materials. *Mater. Rep.* **2022**, *36*, 43–50. [CrossRef]
14. Zheng, X.; Yuan, L. Research status and recent advances in composite humidity control materials. *Chem. Ind. Eng. Prog.* **2020**, *39*, 1378–1388. [CrossRef]
15. Chen, Z.; Qin, M. Theory, preparation and performance of composite heat and humidity control materials. *Archit. Pract.* **2019**, *2*, 10–11. Available online: https://kns.cnki.net/kcms2/article/abstract?v=-PyPURV5YK0JbWU0uvE76XelugL6b80eER5iY682w132Bj_JNLU2-p_LlIbS5oC8GXGfXLNiEFmVyAeh3eg_fGXw6Ggh92oelBdyy6okMzP1ZNq1Ifc1LSl_vjO3CuqmOMbdTX83NrzDZYKarjmjag5ve-UVSWjbZ0sP6RPf9DrPa8rujN_JTQ==&uniplatform=NZKPT&language=CHS (accessed on 25 November 2025).
16. Novais, R.M.; Carvalheiras, J.; Senff, L.; Lacasta, A.M.; Cantalapiedra, I.R.; Giro-Paloma, J.; Seabra, M.P.; Labrincha, J.A. Multifunctional cork–alkali-activated fly ash composites: A sustainable material to enhance buildings' energy and acoustic performance. *Energy Build.* **2020**, *210*, 109739. [CrossRef]
17. Park, J.H.; Kim, Y.U.; Jeon, J.; Yun, B.Y.; Kang, Y.; Kim, S. Analysis of biochar-mortar composite as a humidity control material to improve the building energy and hygrothermal performance. *Sci. Total Environ.* **2021**, *775*, 145784. [CrossRef]
18. Li, B.; Min, F.; Zhang, N.; Ma, J.; Li, Z.; Yao, Z.; Zhang, L. Experimental study on water transfer mechanism of quicklime modified centrifugal dewatering clay. *Constr. Build. Mater.* **2023**, *408*, 133689. [CrossRef]
19. Zhu, Z.; Xiao, P.; Kang, A.; Kou, C.; Chen, J. Preparation of biomass composites with high performance and carbon sequestration from waste wood fibers. *Constr. Build. Mater.* **2023**, *404*, 133295. [CrossRef]
20. Lu, Z.; Zhao, Z.; Wang, M.; Jia, W. Effects of corn stalk fiber content on properties of biomass brick. *Constr. Build. Mater.* **2016**, *127*, 11–17. [CrossRef]
21. Pedreño-Rojas, M.A.; Porras-Amores, C.; Villoria-Sáez, P.; Morales-Conde, M.J.; Flores-Colen, I. Characterization and performance of building composites made from gypsum and woody-biomass ash waste: A product development and application study. *Constr. Build. Mater.* **2024**, *419*, 135435. [CrossRef]
22. Wang, S.; Li, H.; Zou, S.; Liu, L.; Bai, C.; Zhang, G.; Fang, L. Experimental study on durability and acoustic absorption performance of biomass geopolymer-based insulation materials. *Constr. Build. Mater.* **2022**, *361*, 129575. [CrossRef]
23. Bouferra, R.; Bouchehma, A.; Bahammou, Y.; Essaleh, M.; Belhouideg, S.; Lamharrar, A.; Idlimam, A. Experimental study of relative humidity absorbed through capillarity of building material under isothermal conditions: Clay-Kaolin reinforced with Alfa fibers. *Int. Commun. Heat Mass Transf.* **2024**, *154*, 107416. [CrossRef]
24. Wang, M.; Lu, Z. Thermal and moisture coupling performances of poplar fiber inorganic humidity control brick. *Constr. Build. Mater.* **2022**, *348*, 128656. [CrossRef]
25. Castañeda Múnera, J.C.; Gosselin, L. Impact of Density, relative Humidity, and fiber size on hygrothermal properties of barley straw for building envelopes. *Constr. Build. Mater.* **2023**, *409*, 133954. [CrossRef]
26. Palumbo, M.; Lacasta, A.M.; Giraldo, M.P.; Haurie, L.; Correal, E. Bio-based insulation materials and their hygrothermal performance in a building envelope system (ETICS). *Energy Build.* **2018**, *174*, 147–155. [CrossRef]
27. Liuzzi, S.; Rubino, C.; Stefanizzi, P.; Petrella, A.; Boghetich, A.; Casavola, C.; Pappalettera, G. Hygrothermal properties of clayey plasters with olive fibers. *Constr. Build. Mater.* **2018**, *158*, 24–32. [CrossRef]

28. Bahammou, Y.; Kouhila, M.; Tagnamas, Z.; Lamsyehe, H.; Lamharrar, A.; Idlimam, A. Hygroscopic behavior of water absorbed by capillarity and stabilization of a bio-composite building material: Clay reinforced with *Chamarrops humilis* fibers. *Int. Commun. Heat Mass Transf.* **2022**, *135*, 106077. [CrossRef]
29. Li, M.; Luo, F.; Liang, X.; Wang, S.; Gao, X.; Zhang, Z.; Fang, Y. Myristic acid-tetradecanol-capric acid ternary eutectic/SiO_2/MIL-100(Fe) as phase change humidity control material for indoor temperature and humidity control. *J. Energy Storage* **2023**, *74*, 109437. [CrossRef]
30. Izadi, M.; Taghavi, S.F.; Neshat Safavi, S.H.; Afsharpanah, F.; Yaïci, W. Thermal management of shelter building walls by PCM macro-encapsulation in commercial hollow bricks. *Case Stud. Therm. Eng.* **2023**, *47*, 103081. [CrossRef]
31. Al-Shannaq, R.; Farid, M.M.; Ikutegbe, C.A. Methods for the Synthesis of Phase Change Material Microcapsules with Enhanced Thermophysical Properties—A State-of-the-Art Review. *Micro* **2022**, *2*, 426–474. [CrossRef]
32. Sun, Y.; He, F.; Wang, J.; Wang, L.; Fu, B.; Tang, W. Preparation and characterization of fatty acid ternary eutectic mixture/Nano-SiO_2 composite phase change material for building applications. *Appl. Energy* **2024**, *376*, 124221. [CrossRef]
33. Chen, Z.; Qin, M. Preparation and hygrothermal properties of composite phase change humidity control materials. *Appl. Therm. Eng.* **2016**, *98*, 1150–1157. [CrossRef]
34. Wang, X.; Lei, Y.; Chen, Z.; Lei, W. Sepiolite-zeolite powder doped with capric acid phase change microcapsules for temperature-humidity control. *J. Colloid Interface Sci.* **2021**, *595*, 25–34. [CrossRef]
35. Fraine, Y.; Seladji, C.; Aït-Mokhtar, A. Effect of microencapsulation phase change material and diatomite composite filling on hygrothermal performance of sintered hollow bricks. *Build. Environ.* **2019**, *154*, 145–154. [CrossRef]
36. He, X.; Zhang, H.; Qiu, L.; Mao, Z.; Shi, C. Hygrothermal performance of temperature-humidity controlling materials with different compositions. *Energy Build.* **2021**, *236*, 110792. [CrossRef]
37. Yang, Q.; Sun, P.Z.; Fumagalli, L.; Stebunov, Y.V.; Haigh, S.J.; Zhou, Z.W.; Grigorieva, I.V.; Wang, F.C.; Geim, A.K. Capillary condensation under atomic-scale confinement. *Nature* **2020**, *588*, 250–253. [CrossRef] [PubMed]
38. Zhang, S.; Geng, W.; He, X.; Tu, J.; Ma, M.; Duan, L.; Zhang, Q. Humidity sensing performance of mesoporous CoO(OH) synthesized via one-pot hydrothermal method. *Sens. Actuators B Chem.* **2019**, *280*, 46–53. [CrossRef]
39. Lu, H.C.; Kuok, C.H.; Liu, S.H. High-performance humidity control coatings prepared from inorganic wastes. *Constr. Build. Mater.* **2020**, *263*, 120169. [CrossRef]
40. Gao, X.; Asgar, H.; Kuzmenko, I.; Gadikota, G. Architected mesoporous crystalline magnesium silicates with ordered pore structures. *Microporous Mesoporous Mater.* **2021**, *327*, 111381. [CrossRef]
41. Zhou, X.; Jin, H.; Gu, A.; Li, X.; Sun, L.; Mao, P.; Yang, Y.; Ding, S.; Chen, J.; Yun, S. Eco-friendly hierarchical porous palygorskite/wood fiber aerogels with smart indoor humidity control. *J. Clean. Prod.* **2022**, *335*, 130367. [CrossRef]
42. Qin, M.; Hou, P.; Wu, Z.; Wang, J. Precise humidity control materials for autonomous regulation of indoor moisture. *Build. Environ.* **2020**, *169*, 106581. [CrossRef]
43. Hou, P.; Zu, K.; Qin, M.; Cui, S. A novel metal-organic frameworks based humidity pump for indoor moisture control. *Build. Environ.* **2021**, *187*, 107396. [CrossRef]
44. Ge, L.; Feng, Y.; Xue, Y.; Dai, Y.; Wang, R.; Ge, T. Mesoporous Silica-Guided Synthesis of Metal-Organic Framework with Enhanced Water Adsorption Capacity for Smart Indoor Humidity Regulation. *Small Struct.* **2023**, *4*, 2300173. [CrossRef]
45. Liu, X.; Chen, Z.; Yang, G.; Gao, Y. Bioinspired Ant-Nest-Like Hierarchical Porous Material Using $CaCl_2$ as Additive for Smart Indoor Humidity Control. *Ind. Eng. Chem. Res.* **2019**, *58*, 2153–2162. [CrossRef]
46. Yang, Z.; Zhang, W.; Lin, X.; Xiong, Q.; Jiang, Q. Optimization of minor-LiCl-modified gypsum as an effective indoor moisture buffering material for sensitive and long-term humidity control. *Build. Environ.* **2023**, *229*, 109962. [CrossRef]
47. Hung, M.-J.; Lin, Y.-W.; Lin, W.-T.; Lee, W.-H.; Kuo, B.-Y.; Lin, K.-L. Functionalization of mesoporous Al-MCM-41 for indoor humidity control as building humidity conditioning material. *J. Mol. Struct.* **2024**, *1299*, 137024. [CrossRef]
48. Xue, Z.; Wang, J.; Diao, Y.; Hu, W. Study on Preparation and humidity-regulating Capabilities of Vermiculite/Poly(sodium Acrylate-acrylamide) Humidity Controlling Composite. *Materials* **2024**, *17*, 1920. [CrossRef]
49. Yue, L.; Qiu, P.; Zhang, Y.; Miao, J.; Jing, Z. Hydrothermal synthesis of an expanded perlite based lightweight composite for indoor humidity regulating. *Constr. Build. Mater.* **2022**, *351*, 128960. [CrossRef]
50. Liu, X.; Chen, Z.; Yang, G.; Zhang, Z.; Gao, Y. Colorful Wall-Bricks with Superhydrophobic Surfaces for Enhanced Smart Indoor Humidity Control. *ACS Omega* **2019**, *4*, 13896–13901. [CrossRef]
51. Bo, Z.; Meijia, R.; Wenjian, H.; Linying, S.; Erlin, M.; Jun, L. Study on temperature-humidity controlling performance of zeolite-based composite and its application for light quality energy-conserving building envelope. *Alex. Eng. J.* **2023**, *69*, 311–321. [CrossRef]
52. Chehade, S.; Dujardin, N.; Giovannacci, D.; Boudenne, A. Experimental and numerical evaluation of heat and mass transfer of a compressed raw earth block wall. *Case Stud. Constr. Mater.* **2025**, *22*, e04793. [CrossRef]
53. Li, H.; Liu, T.; Peng, Y.; Cai, C.; Liu, L.; Wang, S. Experimental study on enhancing the waterproof and moisture-proof performance of foam open-cell reed-based insulation material. *J. Build. Eng.* **2025**, *103*, 112069. [CrossRef]

54. Zhou, K.; Shi, Z.; Yi, X.-H.; Wang, P.; Li, A.; Wang, C.-C. MOFs helping heritage against environmental threats. *Chin. Chem. Lett.* **2025**, *36*, 110226. [CrossRef]
55. Chen, W.; Yin, Y.; Zhao, X.; Xu, G.; Cao, B.; Ji, Q. Experimental investigation on condensation characteristics of a novel radiant terminal based on sepiolite composite humidity-conditioning coating. *Build. Environ.* **2022**, *223*, 109488. [CrossRef]
56. Zhang, H.; Fu, L.; Wang, X.; Chang, J. Condensation retardation performance of metal-organic framework-based composite humidity-regulating materials in a radiation cooling room. *Build. Environ.* **2023**, *243*, 110630. [CrossRef]
57. Chen, W.; Yin, Y.; Zhao, X.; Fan, F.; Cao, B.; Ji, Q.; Xu, G. Sepiolite based humidity-control coating specially for alleviate the condensation problem of radiant cooling panel. *Energy* **2023**, *272*, 127129. [CrossRef]
58. Zhen, Y.; Zhang, J.; Wang, W.; Li, Y.; Gao, X.; Xue, H.; Liu, X.; Jia, Z.; Xue, Q.; Zhang, J.; et al. Embedded SnO_2/Diatomaceous earth composites for fast humidity sensing and controlling properties. *Sens. Actuators B Chem.* **2020**, *303*, 127137. [CrossRef]
59. Zhan, K.; Zhu, Y.; Yan, J.; Chen, Y. Enhanced-performance relative humidity sensor based on MOF-801 photonic crystals. *Phys. Lett. A* **2020**, *384*, 126678. [CrossRef]
60. Wu, K.; Guan, X.; Hou, Z.; Liu, L.; Zhao, H.; Liu, S.; Fei, T.; Zhang, T. Humidity sensors based on metal organic frameworks derived polyelectrolyte films. *J. Colloid Interface Sci.* **2021**, *602*, 646–653. [CrossRef]
61. Prasetya, N.; Okur, S. Investigation of the free-base Zr-porphyrin MOFs as relative humidity sensors for an indoor setting. *Sens. Actuators A Phys.* **2024**, *377*, 115713. [CrossRef]
62. Wu, K.; He, Q. Highly sensitive low humidity sensor by directional construction of ionic liquid functionalized MOF-303. *Sens. Actuators B Chem.* **2025**, *425*, 136992. [CrossRef]
63. Alsuhaibani, A.M.; Alayyafi, A.A.; Albedair, L.A.; El-Desouky, M.G.; El-Bindary, A.A. Efficient fabrication of a composite sponge for Cr(VI) removal via citric acid cross-linking of metal-organic framework and chitosan: Adsorption isotherm, kinetic studies, and optimization using Box-Behnken design. *Mater. Today Sustain.* **2024**, *26*, 100732. [CrossRef]
64. Zhao, S.; Ma, Y.; Zhou, J. Removal of cadmium ion from wastewater by manganese oxides-loaded sludge biochar. *Desalination Water Treat.* **2024**, *319*, 100563. [CrossRef]
65. Lewis, W.K. The Rate of Drying of Solid Materials. *J. Ind. Eng. Chem.* **1921**, *13*, 427–432. [CrossRef]
66. Stefan, J. Über das Gleichgewicht und Bewegung, insbesondere die Diffusion von Gemischen. Sitzber. *Akad. Wiss. Wien Math.-Naturwiss. Kl. Abt. 2a* **1871**, *63*, 63–124.
67. Buckinghan, E.A. *Studies on the Movement of Soil Moisture*; Bulletin No.: 38; U.S. Department of Agriculture: Washington, DC, USA, 1907.
68. Miller, E.E.; Miller, R.D. Theory of Capillary Flow: I. Practical Implications. *Soil Sci. Soc. Am. J.* **1955**, *19*, 267–271. [CrossRef]
69. Cao, S.; Li, X.; Yang, B. Heat and moisture transfer of building envelopes under dynamic and steady-state operation mode of indoor air conditioning. *J. Build. Eng.* **2021**, *44*, 102683. [CrossRef]
70. Dang, X.; Janssen, H.; Roels, S. Hygrothermal Modelling of one-dimensional Wall Assemblies: Inter-model Validation between WUFI and DELPHIN. *J. Phys. Conf. Ser.* **2023**, *2654*, 012040. [CrossRef]
71. Zeng, L.; Chen, Y.; Cao, C.; Lv, L.; Gao, J.; Li, J.; Zhang, C. Influence of materials' hygric properties on the hygrothermal performance of internal thermal insulation composite systems. *Energy Built Environ.* **2023**, *4*, 315–327. [CrossRef]
72. Belarbi, Y.E.; Ferroukhi, M.Y.; Issaadi, N.; Poullain, P.; Bonnet, S. Assessment of hygrothermal performance of raw earth envelope at overall building scale. *Energy Build.* **2024**, *310*, 114119. [CrossRef]
73. Ferroukhi, M.Y.; Belarbi, R.; Limam, K.; Bosschaerts, W. Experimental validation of a HAM-BES co-simulation approach. *Energy Procedia* **2017**, *139*, 517–523. [CrossRef]
74. Yang, J.; Fu, H.; Qin, M. Evaluation of Different Thermal Models in EnergyPlus for Calculating Moisture Effects on Building Energy Consumption in Different Climate Conditions. *Procedia Eng.* **2015**, *121*, 1635–1641. [CrossRef]
75. Al-Janabi, A.; Kavgic, M.; Mohammadzadeh, A.; Azzouz, A. Comparison of EnergyPlus and IES to model a complex university building using three scenarios: Free-floating, ideal air load system, and detailed. *J. Build. Eng.* **2019**, *22*, 262–280. [CrossRef]
76. Busser, T.; Julien, B.; Amandine, P.; Mickael, P.; Woloszyn, M. Comparison of model numerical predictions of heat and moisture transfer in porous media with experimental observations at material and wall scales: An analysis of recent trends. *Dry. Technol.* **2019**, *37*, 1363–1395. [CrossRef]
77. Defo, M.; Lacasse, M.A.; Laouadi, A. A comparison of hygrothermal simulation results derived from four simulation tools. *J. Build. Phys.* **2021**, *45*, 432–456. [CrossRef]
78. Paepcke, A.; Nicolai, A. Performance analysis of coupled quasi-steady state air flow calculation and dynamic simulation of hygrothermal transport inside porous materials. *Energy Procedia* **2017**, *132*, 759–764. [CrossRef]
79. Luikov, A.V. Heat and Mass Transfer in Capillary-Porous Bodies. In *Advances in Heat Transfer*, 1st ed.; Irvine, T.F., Hartnett, J.P., Eds.; Elsevier: Amsterdam, The Netherlands, 1964; Volume 1, pp. 123–184.
80. De Vries, D.A. The theory of heat and moisture transfer in porous media revisited. *Int. J. Heat Mass Transf.* **1987**, *30*, 1343–1350. [CrossRef]

81. Liu, X.; Chen, Y.; Ge, H.; Fazio, P.; Chen, G. Numerical investigation for thermal performance of exterior walls of residential buildings with moisture transfer in hot summer and cold winter zone of China. *Energy Build.* **2015**, *93*, 259–268. [CrossRef]
82. Yuan, Y.; Tan, L.; Zhao, Z.; Xu, Y.; Bai, M.; Yuan, Y. Pore Network Simulation and Experimental Investigation on Water-Heat Transport Process of Soil Porous Media. *J. Heat Transf.* **2019**, *141*, 102401. [CrossRef]
83. Bianchi Janetti, M.; Janssen, H. Effect of dynamic contact angle variation on spontaneous imbibition in porous materials. *Transp. Porous Media* **2022**, *142*, 493–508. [CrossRef]
84. Fang, Y.; Chen, L.; Wu, L. Transient simulation of coupled heat and moisture transfer through multi-layer walls exposed to future climate in the hot and humid southern China area. *Sustain. Cities Soc.* **2020**, *52*, 101812. [CrossRef]
85. Choi, Y.; Cho, W.; Ozaki, A.; Lee, H. Influence of the moisture driving force of moisture adsorption and desorption on indoor hygrothermal environment and building thermal load. *Energy Build.* **2021**, *253*, 111501. [CrossRef]
86. Pham, S.T.; Chareyre, B.; Tsotsas, E.; Kharaghani, A. Pore network modeling of phase distribution and capillary force evolution during slow drying of particle aggregates. *Powder Technol.* **2022**, *407*, 117627. [CrossRef]
87. Zhao, C.; Zhou, W.; Chen, X.; Wu, Q.; Ma, G.; Wang, Q. Mesoscopic analysis of heat and moisture coupled transfer in concrete considering phase change under frost action. *J. Build. Eng.* **2022**, *57*, 104888. [CrossRef]
88. Sadeghi, R.; Shadloo, M.S.; Hopp-Hirschler, M.; Hadjadj, A.; Nieken, U. Three-dimensional lattice Boltzmann simulations of high density ratio two-phase flows in porous media. *Comput. Math. Appl.* **2018**, *75*, 2445–2465. [CrossRef]
89. Kuenzel, H. Simultaneous Heat and Moisture Transport in Building Components. One- and Two-Dimensional Calculation Using Simple Parameters. In *Research Reports*; Fraunhofer IBP, Ed.; Fraunhofer IBP: Stuttgart, Germany, 1995.
90. Kaoutari, T.; Louahlia, H. Experimental and numerical investigations on the thermal and moisture transfer in green dual layer wall for building. *Case Stud. Therm. Eng.* **2024**, *53*, 103946. [CrossRef]
91. Alioua, T.; Agoudjil, B.; Boudenne, A.; Benzarti, K. Sensitivity analysis of transient heat and moisture transfer in a bio-based date palm concrete wall. *Build. Environ.* **2021**, *202*, 108019. [CrossRef]
92. Bart, M.; Sophie, M.; Yacine, A.O.; Lanos, C. Transient hygrothermal modelling of coated hemp-concrete walls. *Eur. J. Environ. Civ. Eng.* **2014**, *18*, 927–944. [CrossRef]
93. Dong, W.; Chen, Y.; Bao, Y.; Fang, A. A validation of dynamic hygrothermal model with coupled heat and moisture transfer in porous building materials and envelopes. *J. Build. Eng.* **2020**, *32*, 101484. [CrossRef]
94. Janssen, H. A comment on "A validation of dynamic hygrothermal model with coupled heat and moisture transfer in porous building materials and envelopes". *J. Build. Eng.* **2022**, *47*, 103835. [CrossRef]
95. Bennai, F.; Abahri, K.; Belarbi, R. Contribution to the Modelling of Coupled Heat and Mass Transfers on 3D Real Structure of Heterogeneous Building Materials: Application to Hemp Concrete. *Transp. Porous Media* **2020**, *133*, 333–356. [CrossRef]
96. Dahanni, H.; Rima, A.; Abahri, K.; Hachem, C.E.; Assoum, H. Impact of the 3D morphology on the hygro-thermal transfer of hygroscopic materials: Application to spruce wood. *J. Phys. Conf. Ser.* **2021**, *2069*, 012008. [CrossRef]
97. Galy, T.; Mu, D.; Marszewski, M.; Pilon, L. Computer-generated mesoporous materials and associated structural characterization. *Comput. Mater. Sci.* **2019**, *157*, 156–167. [CrossRef]
98. Nasir, M.; Li, Z.; Mahardika, M.A.; Wang, W.; She, Y.; Wang, K.; Patmonoaji, A.; Matsushita, S.; Suekane, T. Pore-scale investigation of wettability effects on drying process of three-dimensional porous medium. *Int. Commun. Heat Mass Transf.* **2023**, *140*, 106527. [CrossRef]
99. Fan, C.; Liu, R.; Liao, Y. Archetype Identification and Energy Consumption Prediction for Old Residential Buildings Based on Multi-Source Datasets. *Buildings* **2025**, *15*, 2573. [CrossRef]
100. Zhou, P.; Qi, Y.; Yang, Q.; Chang, Y. Neural topic modeling of machine learning applications in building: Key topics, algorithms, and evolution patterns. *Autom. Constr.* **2025**, *170*, 105890. [CrossRef]
101. Li, B.; Yong, J.C.E.; Yu, L.J.; Olugu, E.U.; Yang, X.; Zhang, Z.; Muhieldeen, M.W. A review of research on intelligent technology in building air conditioning system optimisation. *Int. J. Refrig.* **2025**, *176*, 205–225. [CrossRef]
102. Alexakis, K.; Benekis, V.; Kokkinakos, P.; Askounis, D. Genetic algorithm-based multi-objective optimisation for energy-efficient building retrofitting: A systematic review. *Energy Build.* **2025**, *328*, 115216. [CrossRef]

MDPI AG
Grosspeteranlage 5
4052 Basel
Switzerland
Tel.: +41 61 683 77 34

Buildings Editorial Office
E-mail: buildings@mdpi.com
www.mdpi.com/journal/buildings

www.ingramcontent.com/pod-product-compliance
Lightning Source LLC
Chambersburg PA
CBHW061623210926
51850CB00009B/2415